Study Guide to Accompany

Nutrition *for* Foodservice *and* Culinary Professionals

Seventh Edition

Karen Eich Drummond
Ed.D., R.D., L.D.N., F.A.D.A., F.M.P.

Lisa M. Brefere
C.E.C., A.A.C.

JOHN WILEY & SONS, INC.

This book is printed on acid-free paper. ♾

Published by John Wiley & Sons, Inc., Hoboken, New Jersey
Published simultaneously in Canada

For general information on our other products and services or for technical support, please contact our Customer Care Department within the United States at (800) 762-2974, outside the United States at (317) 572-3993 or fax (317) 572-4002.

Wiley also publishes its books in a variety of electronic formats. Some content that appears in print may not be available in electronic books. For more information about Wiley products, visit our website at www.wiley.com.

Library of Congress Cataloging-in-Publication Data:

ISBN: 978-0-470-28547-3

Printed in the United States of America

10 9 8 7 6 5 4 3

TABLE OF CONTENTS

PREFACE

This Study Guide to accompany *Nutrition for Foodservice and Culinary Professionals, Seventh Edition* by Karen Eich Drummond and Lisa M. Brefere was prepared to help you learn and apply the concepts in each chapter. To this end we have included the following materials for each chapter:

1. Learning Objectives
2. Chapter Outline
3. Nutrition Web Explorer
4. Chapter Review Quiz
5. Student Worksheets

The "Learning Objectives" indicate what you can expect to learn from this course, and are designed to help you organize your studying and concentrate on important topics and explanations.

The "Chapter Outline" summarizes the key ideas and concepts from the chapter.

"Nutrition Web Explorer" includes exercises that use Internet websites to help you apply concepts learned in the chapter.

The "Chapter Review Quiz" uses matching of key chapter terms, true/false, multiple choice, and fill-in questions to help you review the chapter and prepare for tests.

The "Student Worksheets" are fun activities that make you apply key concepts from each chapter.

PowerPoint Presentations for each chapter can be downloaded from the book's website at www.wiley.com/college/drummond.

You will also find on the website over 200 healthy recipes organized by chapter and topic.

CHAPTER 1 INTRODUCTION TO NUTRITION

LEARNING OBJECTIVES

Upon completion of the chapter, the student should be able to:

1. Identify factors that influence food selection
2. Define *nutrition, kilocalorie, nutrient,* and *nutrient density*
3. Identify the classes of nutrients and their characteristics
4. Describe four characteristics of a nutritious diet
5. Define Dietary Reference Intakes and explain their function
6. Compare the EAR, RDA, AI, and UL
7. Describe the processes of digestion, absorption, and metabolism
8. Explain how the digestive system works
9. Distinguish between whole, processed, and organic foods
10. Compare how a meat-based or plant-based diet impact the environment

CHAPTER OUTLINE

1. **Understand factors influencing food selection**

   Many factors influence what you eat: flavor, other aspects of food (cost, convenience, availability, familiarity, nutrition), demographics, culture and religion, health, social and emotional influences, food industry and the media, and environmental concerns.

   Flavor is a combination of all five senses: taste, smell, touch, sight, and sound.

   - Taste comes from 10,000 taste buds found on the tongue, cheeks, throat, and the roof of the mouth. Taste buds are most numerous in children under 6. Taste buds for each sensation (sweet, salt, sour, bitter, and umami) are scattered throughout the mouth.
   - Umami differs from the traditional tastes by providing a savory, sometimes meaty sensation. The umami taste receptor is very sensitive to glutamate, which occurs naturally in foods such as seafood and seaweed (also is in MSG).
   - Your ability to identify the flavors of specific food requires smell, meaning the special cells high up in your nose that detect odors.
   - All foods have texture that you sense when you eat. Food appearance or presentation strongly influences which foods you choose to eat.
   - To some extent, what you smell or taste is genetically determined.

   Food cost is a major consideration, along with convenience, food availability and familiarity, and nutritional content.

   Demographics factors include age, gender, educational level, income, and cultural background. Older adults are often more nutritional-minded, as well as people with higher incomes and educational levels.

   Culture strongly influences the eating habits of its members — norms about which foods are eaten, how often foods are eaten, what foods are eaten together, etc. For many people, religion affects their day-to-day food choices.

Heath status, social and emotional influences, the food industry and the media, and environmental concerns also may influence what you eat.

2. **Explore concepts of nutrition**

   Nutrition is a science that studies nutrients and other substances in foods and in the body and how these nutrients relate to health and disease. Nutrition also explores why you choose particular foods and the type of diet you eat.

   Nutrients are nourishing substances in food that provide energy and promote the growth and maintenance of your body.

3. **Understand kilocalories and how many you need**

   Kilocalories are a measure of the energy in food. A kilocalorie, also called a Calorie, raises the temperature of 1 kilogram of water 1 degree Celsius. One kilocalorie contains 1,000 kcalories. The media uses the term "calorie" when the correct term is kilocalorie.

   The number of kcalories you need is based on:

   - basal metabolism (about 2/3 of energy expended for individuals who are not very active)
   - physical activity
   - thermic effect or specific dynamic action of food (for every 100 kcalories you eat, about 5 to 10 kcalories are used for digestion, absorption, and metabolism)

   Your **basal metabolic rate** depends on factors such as:

   - Gender (males have higher BMR due to more muscle tissue)
   - Age (BMR decreases as we age)
   - Growth (BMR is higher during growth)
   - Height (tall people have more body surface and higher BMR)
   - Temperature (BMR increases in both hot and cold environments)
   - Fever and stress (both increase BMR)
   - Exercise (increases BMR)
   - Smoking and caffeine (increase BMR)
   - Sleep (BMR is at its lowest)

4. **Explore the 6 classes of nutrients**

   Nutrients provide energy, promote the growth and maintenance of the body, and/or regulate body processes. There are about 50 total nutrients that are categorized into these 6 categories.

   - **Carbohydrates** – A large class of nutrients, including sugars, starch, and fibers, that functions as the body's primary source of energy.
   - **Fats (Lipids)** – A group of fatty substances, including triglycerides and cholesterol, that are soluble in fat, not water, and that provide a rich source of energy and structure to cells.
   - **Proteins** – Major structural parts of the body's cells that are made of nitrogen-containing amino acids assembled in chains, particularly rich in animal foods.

- **Vitamins** – Noncaloric, organic nutrients found in a wide variety of foods that are essential in small quantities to regulate body processes, maintain the body, and allow growth and reproduction.
- **Minerals** – Noncaloric, inorganic chemical substances found in a wide variety of foods; needed to regulate body processes, maintain the body, and allow growth and reproduction.
- **Water** plays a vital role in all bodily processes and makes up just over half of your body's weight. It supplies the medium in which various chemical changes of the body occur and aids digestion and absorption, circulation of blood, and lubrication of body joints. (Inorganic)

Most foods provide a mix of nutrients and that food contains more than just nutrients (may contain colorings, flavorings, caffeine, phytochemicals, etc.).

Only carbohydrate, lipids and proteins supply kcalories: carbohydrates and proteins at 4 kcalories/gram and lipids at 9 kcalories per gram. Vitamins and minerals do not contain kcalories.

Organic compounds contain carbon and include carbohydrates, lipids, proteins, and minerals. **Inorganic compounds**, such as minerals and water, don't contain carbon.

"You are what you eat." The nutrients you eat are in your body—water is about 60% of weight, protein is 15%, and so on.

Essential nutrients either cannot be made in the body or cannot be made in the quantities needed —we must therefore obtain them from food. Examples include glucose, vitamins, minerals, water, some lipids, and some parts of protein.

Explain the concept of nutrient density: a measure of the nutrients provides in a food per kcalorie of that food. Give examples of foods with high and low nutrient density. Explain empty-kcalorie foods.

5. **Understand the four characteristics of a nutritious diet**
   - **Adequate diet** – provides enough kcalories, essential nutrients, and fiber to keep a person healthy.
   - **Balanced diet** – foods are chosen to provide kcalories, essential nutrients, and fiber in the right proportions.
   - **Moderate diet** – avoids excess amounts of kcalories or any particular food or nutrient.
   - **Varied diet** – a wide selection of foods are chosen to get necessary nutrients.

6. **The Dietary Reference Intakes (DRI) are nutrient standards that include 4 lists of values for dietary nutrient intakes of healthy Americans and Canadians**
   The DRI include:
   - **Estimated Average Requirement (EAR)** – dietary intake value sufficient to meet the requirement of half the healthy individuals in a group
   - **Recommended Dietary Allowance (RDA)** – dietary intake value sufficient to meet the nutrient requirements of 97–98% of all healthy individuals in a group

- **Adequate Intake (AI)** – dietary intake used when there is not enough evidence to develop a RDA
- **Tolerable Upper Intake Level (UL)** – maximum intake level above which risk of toxicity would increase
- **Estimated Energy Requirement (EER)** – dietary energy intake measured in kcalories that is needed to maintain energy balance in a healthy adult—there is no RDA or UL for kcalories.

DRI vary depending on age, gender, pregnancy, and lactation. They are designed to help healthy people maintain health and prevent disease.

The DRIs are used to assess dietary intakes as well as to plan diets. The RDA and AI are useful in planning diets for individuals.

Acceptable Macronutrient Distribution Ranges (AMDR) is the range of intakes associated with reduced risk of chronic disease while providing adequate intake.

Acceptable Macronutrient Distribution Ranges (% of total kcal)

Age	Carbohydrate	Fat	Protein
1–3 years	45–65%	30–40%	5–20%
4–18 years	45–65%	25–35%	10–30%
Over 18 years	45–65%	20–35%	10–35%

7. **Understand what happens when you eat: digestion, absorption, and metabolism**

Digestion is the process by which food is broken down into its components in the mouth, stomach, and small intestine with the help of digestive enzymes.

Absorption is the passage of digested nutrients through the walls of the intestines or stomach into the body's cells. Nutrients are then transported through the body via the blood or lymph system.

Metabolism refers to all the chemical processes by which nutrients are used to support life—including the processes by which body tissues and substances are built (**anabolism**) and processes by which large molecules are converted to simpler ones (**catabolism**).

The **gastrointestinal tract** is a hollow tube running down the middle of the body in which digestion of food and absorption of nutrients take place.

What happens when you eat:

- 32 permanent teeth grind and break down the food.
- Saliva contains important digestive **enzymes** and lubricates the food so it can pass down the esophagus.
- Tongue moves food around during chewing and rolls the food into a **bolus** (or ball) to be swallowed.

- The **pharynx** connects the mouth and nasal cavities to the esophagus and the air tubes to the lungs.
- **Epiglottis** covers air tubes during swallowing and food enters **esophagus** (muscular tube that leads to stomach). **Peristalsis** (involuntary rhythmic contractions of muscles) occurs in esophagus and helps break up food.
- Food moves through the lower esophageal or cardiac sphincter into the stomach. Stomach holds 4 cups of food and is lined with mucous membrane that makes hydrochloric acid and an enzyme to break down protein.
- **Hydrochloric acid** aids in protein digestion, destroys harmful bacteria, and increases calcium and iron absorption.
- Stomach functions like a holding tank. It is a J-shaped muscular sac that holds about 4 cups of food when full. It takes about 1 1/2–4 hours for it to empty **chyme** into small intestine. Fatty foods take the longest to digest. Mostly alcohol is absorbed here.
- **Small intestine** is 15–30 feet long and has 3 parts: **duodenum, jejunum, and ileum**. It produces digestive enzymes and receives **bile** from the gall bladder (bile is made in the liver, as well as digestive enzymes from the pancreas. Bile is necessary for fat digestion.
- Most nutrients pass through the **villi** (and **microvilli**) of the duodenum into either the blood or lymph vessels where they are transported to the liver and to the cells of the body. Most digestion is completed in the first half of the small intestine. Food is in the small intestine for about 7–8 hours.
- **Large intestine (colon**) is 4 to 5 feet long and ends at rectum. Large intestine receives and stores the waste products of digestion, and reabsorbs water and some minerals. Food is in the large intestine about 18 to 24 hours.
- **Ulcers** are a common digestive problem affecting the duodenum or stomach. A peptic ulcer is a sore on the lining of the stomach or duodenum. Causes include bacterial infection or long-term use of medications such as aspirin. Taking antibiotics, quitting smoking, limiting consumption of caffeine and alcohol and reducing stress can speed healing.
- Body waste is expelled from the **rectum** through the **anus**.

8. **Food Facts: Food Basics**

Whole foods are foods as we get them from nature (for example, eggs, fruits, vegetables)

Processed foods have been prepared using a certain procedure: cooking, freezing, canning, dehydrating, milling, culturing with bacteria, or adding nutrients.

An enriched food is one in which nutrients are added to it to replace the same nutrients that were lost in processing. A fortified food has nutrients added to it that were not present originally. ex. Milk

Organic foods are foods that have been grown without using most synthetic pesticides, fertilizers made with synthetic ingredients, antibiotics, or hormones, and without genetic engineering or irradiation. Food labeling requirements include the following.

- Foods labeled "100% organic" must contain only organically produced ingredients (excluding water and salt).
- Foods labeled "organic" must consist of at least 95% organically produced ingredients (excluding water and salt).

- Foods labeled "made with organic ingredients" must contain at least 70% organic ingredients.
- Processed foods that contain less than 70% organic ingredients cannot use the term "organic" anywhere on the display panel of the food label.

9. **Hot Topic: How the American Diet Impacts the Environment and How Restaurants Are Going Green**

Producing large quantities of meat in America uses many resources and has serious environmental consequences, such as losing forests for pastureland, creating air and water pollution, and requiring enormous amounts of water, fuel, fertilizers, and pesticides to grow feed.

Sustainable agriculture produces abundant food without depleting the earth's resources or polluting its environment. Sustainable practices lend themselves to smaller, family-scale farms.

Many restaurants are involved in the green movement by, for example, purchasing locally grown and organic foods, saving energy, buying tableware and other products made of recycled and renewable materials, buying nontoxic cleaning supplies, saving water, and recycling.

NUTRITION WEB EXPLORER

U.S. Government Healthfinder: www.healthfinder.gov
This government site can help you find information on virtually any health topic. On the home page, click on "H" under "Prevention and Wellness." Next, click on heart disease. Using the links, find 5 way to reduce your risk of heart disease and write below.

National Agricultural Library: www.nutrition.gov
From this government site, you can access many nutrition topics right from the home page. Click on "In the News." Then, click on "Does Meal Frequency Affect Your Health?" and summarize this article in one paragraph below.

National Organic Program: www.ams.usda.gov/nop
Visit the website for the National Organic Program to find out if a "natural" food can also be labeled as "organic." Write your answer below.

Center for Science in the Public Interest: Eating Green: www.cspinet.org/EatingGreen
Click on "Eating Green Calculator" and fill in how much in the way of animal products you eat each week. Then click on "Calculate Impact" and find out the environmental impact of your eating habits. Also use the "Score Your Diet" tool to show how your diet scores on nutrition, the environment, and animal welfare. Summarize your results below.

Center for Young Womens' Health: http://www.youngwomenshealth.org/college101.html
Read College Eating and Fitness 101. List 5 suggestions they make to help you not gain the Freshman 15.

Alcohol Calorie Calculator:
http://www.collegedrinkingprevention.gov/CollegeStudents/calculator/alcoholcalc.aspx
Fill in the "Average Drinks per Week" column and then press "Compute." You will see how many calories you take in each month and in one year from alcoholic beverages. Summarize your results below.

CHAPTER REVIEW QUIZ

Key Terms: Matching

1. Taste buds
2. Basal metabolism
3. Thermic effect of food
4. Energy-yielding nutrients
5. Micronutrient
6. Macronutrient
7. Organic
8. Essential nutrients
9. Metabolism
10. Anabolism
11. Catabolism
12. Bolus
13. Epiglottis
14. Peristalsis
15. Chyme
16. Bile

a. A substance made by the liver that is stored in the gallbladder and released when fat enters the small intestines that are involved in absorption
b. A ball of chewed food that travels from the mouth through the esophagus to the stomach
c. Nutrients that can be burned as fuel to provide energy for the body
d. In chemistry, any compound that contains carbon
e. Clusters of cells found on the tongue, cheeks, throat, and roof of the mouth
f. All the chemical processes by which nutrients are used to support life
g. Nutrients needed by the body, including carbohydrates, lipids, and proteins
h. Semi-liquid mixture in the stomach with partially digested food and stomach secretions
i. The metabolic process by which body tissues and substances are built
j. Flap that covers the air tubes to the lungs so that food does not enter the lungs during swallowing
k. The energy needed to digest and absorb food
l. Nutrients that cannot be made in the body so we must obtain them from food
m. Nutrients needed by the body, including vitamins and minerals
n. Metabolic process where large, complex molecules are converted to simpler ones
o. The minimum energy needed by the body for vital functions
p. Involuntary muscular contraction that forces food through the digestive system

True/False

1. One's age, educational level, and income can affect the foods they eat.
   a. True b. False

2. The number of kcalories you need is based on three factors: your basal metabolism, the thermic effect of foods, and the type of macronutrients consumed.
   a. True b. False

3. Protein is considered an energy-yielding nutrient because it can be burned as fuel to provide energy for the body.
   a. True b. False

4. Although fiber can't be broken down or digested in the body, it is a necessary for healthy body function.
   a. True b. False

5. Minerals and vitamins do not provide energy.
   a. True b. False

6. Carbohydrate is the main structural component of all the body's cells.
   a. True b. False

7. Water is the most plentiful nutrient in the human body.
   a. True b. False

8. Cholesterol and other types of lipids provide structure to the body's cells.
   a. True b. False

9. The DRIs vary depending on age, gender, and health status.
   a. True b. False

10. One's EER depends on age, gender, weight, height, and activity level.
    a. True b. False

11. Enzymes in the pharynx start the digestion of carbohydrate.
    a. True b. False

12. Hydrochloric acid aids in protein digestion, and increases the ability of calcium and iron to be absorbed.
    a. True b. False

13. Bile is produced by the gall bladder but stored in the liver to aid in fat digestion.
    a. True b. False

14. The order of sections of the small intestine is: duodenum, ileum, jejunum.
   a. True   (b.) False

15. The anus stores the waste products until they are released as solid feces through the rectum.
   (a.) True   b. False

Fill in the Blank

1. Triglycerides is a major form of lipids that provide energy for the body, as well as a way to store energy as fat.

2. Protein is the macronutrient that accounts for 15% of body weight.

3. Kids who drink too much soda can be at risk for getting too little Calcium.

4. Nutrients are absorbed into either the blood or Lymph, two fluids that circulate throughout the body, delivering nutrients and picking up waste.

5. The Stomach makes enzymes that break down protein, makes hydrochloric acid, and churns and mixes food.

6. bile is produced by the liver and contains partially digested food and stomach secretions.

7. Minerals are inorganic chemical substances that regulate body processes and allow growth and reproduction without any energy.

8. The Nutrient Density of a food depends on the amount of calories in comparison with nutrients it contains.

9. Water is an inorganic nutrient that plays a vital role in all bodily processes and makes up just over half of the body's weight.

10. RDA is the dietary intake value that is sufficient to meet the nutrient requirements of 97 to 98 percent all healthy individuals.

STUDENT WORKSHEET 1-1

TOPIC: Factors Influencing What You Eat

Why do you eat the foods you commonly eat every day? Describe how at least 6 of the following factors influence what you eat/drink for meals or snacks.

Flavor
Cost
Convenience
Nutrition and Health
Culture
Religion
Food industry/media
Peers
Environmental concerns

STUDENT WORKSHEET 1-2

TOPIC: Dietary Reference Intakes: Terms

1. What does each letter in these acronyms stand for?

   DRI Dietary Reference Intakes

   EAR Estimated Average Requirement

   RDA Recommended Dietary Allowance

   AI Adequate Intake

   UL Tolerable Upper Intake Level

   EER Estimated energy requirement

   AMDR Acceptable Macronutrient Distribution Ranges

2. Match the Dietary Reference Intakes values on the left with their definition.

Dietary Reference

C 1. EAR

E 2. RDA

B 3. AI

F 4. UL

D 5. EER

A 6. AMDR

Definition

A. Range of intakes for a particular nutrient that is associated with reduced risk of chronic disease while providing adequate intake.

B. Dietary intake value used when a RDA can't be developed.

C. Dietary intake value estimated to meet the requirement of half the healthy individuals in a group.

D. Dietary energy intake needed to maintain energy balance in a healthy adult.

E. Dietary intake value sufficient to meet the nutrient requirements of 97–98% of all healthy individuals in a group.

F. Maximum intake level above which risk of toxicity increases.

STUDENT WORKSHEET 1-3

TOPIC: Dietary Reference Intakes: Applications

1. Use Appendix B ("Dietary Reference Intakes") to find out the following. Keep in mind that nutrients are measured in different ways. For example: grams (commonly abbreviated as "g"—about the weight of a paperclip), milligrams ("mg"—1/1,000th of a gram), and micrograms ("ug"—1/1,000,000th of a gram).

   A. How much vitamin C do you need each day?

   65 mg/day

   B. Is your requirement for vitamin C a RDA or AI?

   RDA

   C. What is the requirement for calcium for a 25-year-old woman who is pregnant?

   1000 mg/day

   D. Is the calcium requirement a RDA or AI?

   AI

   E. Who needs more iron: a 15-year-old boy or girl?

   Girl

   F. What is the Tolerable Upper Intake Level for vitamin E for a 75-year-old woman?

   1000 mg/day

   G. What is the RDA for carbohydrate, fat, and protein for a 20 year old?

   H. What is the AI for total fiber for a woman during pregnancy?

   28g/day

   I. What is the AI for total fat for a 30-year-old female?

CHAPTER 2 USING DIETARY RECOMMENDATIONS, FOOD GUIDES, AND FOOD LABELS TO PLAN MENUS

LEARNING OBJECTIVES

1. Discuss the Dietary Guidelines for Americans with regard to adequate nutrients within kcalorie needs, weight management, physical activity, foods groups to encourage, fat, carbohydrates, sodium and potassium, alcoholic beverages, and food safety
2. Recommend ways to implement each Dietary Guideline
3. Describe each food group in MyPyramid including subgroups as appropriate
4. Explain the concept of discretionary kcalories
5. Gives examples of portion sizes from each food group
6. Describe how MyPyramid illustrates variety, proportionality, and moderation
7. Plan menus using MyPyramid
8. List the information required on a food label
9. Read and interpret information from the Nutrition Facts label
10. Distinguish between a nutrient claim and a health claim
11. Explain how an "A" health claim differs from those ranked "B", "C", or "D"
12. Discuss the relationship between portion size on food labels and portions in MyPyramid

CHAPTER OUTLINE

1. **Understand dietary recommendations and food guides**

   Whereas dietary recommendations discuss specific foods to eat for optimum health, food guides (such as MyPyramid) tell us the amounts of foods we need to eat to have a nutritionally adequate diet. Food guides are based on current dietary recommendations, the nutrient content of foods, and the eating habits of the targeted population. MyPyramid is based on the Dietary Guidelines for Americans and nutrient recommendations.

2. **Explore the Dietary Guidelines for Americans**

   Adequate Nutrients Within Kcalorie Needs

   Key Recommendations

   - Meet recommended intakes within energy needs by adopting a balanced eating pattern such as that in MyPyramid. This food guide is designed to integrate dietary recommendations into a healthy way to eat. MyPyramid differs in important ways from common food consumption patterns in the United States. In general, MyPyramid recommends:
     - More dark green vegetables, orange vegetables, legumes, fruits, whole grains, and low-fat milk and milk products
     - Less refined grains, total fats (especially cholesterol, and saturated and trans fats), added sugar, and kcalories
   - Consume a variety of nutrient-dense foods and beverages within and among the basic food groups while choosing foods that limit the intake of saturated and trans fats, cholesterol, added sugars, salt, and alcohol.

Weight Management

Key Recommendations

- To maintain body weight in a healthy range, balance kcalories from foods and beverages with kcalories expended.
- To prevent gradual weight gain over time, make small decreases in food and beverage kcalories and increase physical activity.
- If you need to lose weight, aim for a slow, steady weight loss by decreasing kcalorie intake while maintaining an adequate nutrient intake and increasing physical activity. A reduction of 500 kcalories per day is often needed and will result in weight loss of about 1 pound a week. When it comes to losing weight, it is kcalories that count—not the proportions of fat, carbohydrates, and protein in the diet.

Physical Activity

Key Recommendations

- Engage in regular physical activity and reduce sedentary activities to promote health, psychological well-being, and a healthy body weight.
    - To reduce the risk of chronic disease in adulthood, engage in at least 30 minutes of moderate-intensity physical activity, above usual activity, on most days of the week.
    - For most people, greater health benefits can be obtained by engaging in physical activity of more vigorous intensity or longer duration.
    - To help manage body weight and prevent gradual, unhealthy body weight gain in adulthood, engage in approximately 60 minutes of moderate- to vigorous-intensity activity on most days of the week while not exceeding caloric intake requirements.
    - To sustain weight loss in adulthood: participate in at least 60 to 90 minutes of daily moderate-intensity physical activity while not exceeding caloric intake requirements. Some people may need to consult with a healthcare provider before participating in this level of activity.
- Achieve physical fitness by including cardiovascular conditioning, stretching exercises for flexibility, and resistance exercises or calisthenics for muscle strength and endurance.

Food Groups to Encourage

Key Recommendations

- Consume a sufficient amount of fruits and vegetables while staying within energy needs. Two cups of fruit and 2 1/2 cups of vegetables per day are recommended for a reference 2,000 kcalorie intake.
- Choose a variety of fruits and vegetables each day. In particular, select from all five vegetable subgroups (dark green, orange, legumes, starchy vegetables, and other vegetables) several times a week.
- Consume 3 or more ounce-equivalents of whole-grain products per day, with the rest of the recommended grains coming from enriched or whole-grain products. In general, at least half the grains should be whole grains.
- Consume 3 cups per day of fat-free or low-fat milk or equivalent milk products.

Fats

Key Recommendations

- Consume less than 10% of kcalories from saturated fatty acids and less than 300 mg/day of cholesterol, and keep trans fatty acid consumption as low as possible.
- Keep total fat intake between 20 to 35% of kcalories, with most fats coming from sources of polyunsaturated and monounsaturated fatty acid, such as fish, nuts, and vegetable oils.
- When selecting and preparing meat, poultry, dry beans, and milk or milk products, make choices that are lean, low-fat, or fat-free.
- Limit intake of fats and oils high in saturated and/or trans fatty acids, and choose products low in such fats and oils.

Carbohydrates

Key Recommendations

- Choose fiber-rich fruits, vegetables, and whole grains often.
- Choose and prepare foods and beverages with little added sugars or caloric sweeteners, such as amounts suggested by the USDA Food Guide.
- Reduce the incidence of dental caries by practicing good oral hygiene and consuming sugar- and starch-containing foods and beverages less frequently.

Sodium and Potassium

Key Recommendations

- Consume less than 2,300 mg (about 1 teaspoon of salt) of sodium per day.
- Choose and prepare foods with little salt.
- Eat potassium-rich foods such as fruits and vegetables. (Potatoes and Bananas)

Alcoholic Beverages

Key Recommendations

- Those who choose to drink alcoholic beverages should do so sensibly and in moderation—defined as the consumption of up to one drink per day for women and up to two drinks per day for men.
- Alcoholic beverages should not be consumed by some individuals, including those who cannot restrict their alcohol intake, women of childbearing age who may become pregnant, pregnant and lactating women, children and adolescents, individuals taking medications that can interact with alcohol, and those with specific medical conditions.
- Alcoholic beverages should be avoided by individuals engaging in activities that require attention, skill, or coordination, such as driving or operating machinery.

Food Safety

3. **Use MyPyramid**

Principles it illustrates:

1. One size doesn't fit all – 12 MyPyramids range from 1,000 to 3,200 kcalories.
2. Activity – steps and person climbing.
3. Moderation – narrowing of each food group from bottom to top, wider base stands for foods with little or no solid fats or added sugars.

4. Proportionality – shown by the different widths of the food group bands.
5. Variety – symbolized by the 6 color bands representing the 5 food groups plus oils.
6. Gradual improvement – "Steps to a Healthier You"

Number of servings for 2,000 kcal/day:

6 ounces of grains
2.5 cups of vegetables
2 cups of fruit
3 cups of milk
5.5 ounces of lean meat, beans, or equivalent
27 grams of oils
267 discretionary kcalories

Grain Group

Two subgroups: whole grains (whole wheat bread, oatmeal, brown rice, bulgur) refined grains (white bread, white rice)

All age groups should eat at least half the grains as whole grains.

Serving size: 1 slice of bread, 1 cup of ready-to-eat cereal, 1 small muffin, ½ cup of cooked rice, cooked pasta, or cooked cereal.

Sources of several B vitamins (thiamin, riboflavin, niacin, and folate) and minerals (iron and copper). Whole grains are also good sources of dietary fiber, magnesium, and selenium.

Vegetable Group

Five subgroups:
- Dark green veggies
- Orange veggies
- Dry beans and peas
- Starchy vegetables
- Other vegetables

A weekly intake of specific amounts from each subgroup is recommended.

1 Cup of vegetables = 1 cup of raw or cooked vegetables, 1 cup vegetable juice, 2 cups of raw leafy greens.

Vegetables are low in kcalories and fat, and are important sources of fiber, vitamin A, vitamin C, vitamin E, folate, magnesium, and potassium.

Fruit Group

1 cup of fruit = 1 cup 100% fruit juice, 1 cup fruit, ½ cup dried fruit, 1 small apple, 1 large banana, 32 seedless grapes, 1 medium pear, 1 large orange.

Fruits are low in kcalories , fat, and sodium, and they are important sources of vitamin C, potassium, folate, and fiber.

Milk Group

1 cup of milk = 1 cup yogurt, 1 1/2 ounces of natural cheese, 2 ounces of processed cheese. Make choices that are low in fat or fat-free.

Meat and Beans Group

Meat and poultry choices should be lean or low-fat.

Fish, nuts, and seeds contain healthy oils, so you can choose these.
Dry beans and peas can be counted either in the vegetable group or the meat and beans group.
1 ounce-equivalent = 1 ounce of meat, poultry, or fish; ½ cup cooked dry beans; 1 egg; 1 tablespoon of peanut butter; or ½ ounce of nuts or seeds.
The meat and bean group supplies many nutrients: protein, B vitamins, vitamin E, iron, zinc, magnesium.
Some food choices are high in saturated fat and cholesterol.

Oils

Includes fats that are liquid at room temperature, such as vegetable oils.
Most of the fats you eat should be monounsaturated or polyunsaturated.

Discretionary Kcalories

Includes most solid fats and all added sugars, even if they are part of your selections from the grains group, lean meat group, or another food group.
Examples of discretionary kcalories: fat in cheese, fat in poultry skin or most luncheon meats, sugar added to fruit drinks, added fats and/or sugars in pies, cookies, etc.
At 2,000 kcalories per day, you have 265 discretionary kcalories.

Physical Activity

At a minimum, do moderate-intensity activity for 30 minutes most days or preferably every day. This is in addition to your usual daily activities. About 60 minutes a day may be needed to prevent weight gain.

Planning Menus Using MyPyramid

1. Does a day's menu on the average provide at least the number of servings required from each of the major food groups for a 2,000-kcalorie diet?
2. Are most of the menu items nutrient-dense (without solid fat or sugars added)?
3. Does the menu have whole-grain breads, etc. at each meal?
4. Are most meat and poultry items lean?
5. Are fish, beans, and other meat alternates available?
6. Does the menu include servings from each of the vegetable subgroups: dark orange, green, beans, starchy, and other?
7. Do most veggies and fruits have their skins and seeds?
8. Are there more choices for fresh, canned, or dried fruit than for fruit juices?
9. Are low-fat or fat-free milk and other dairy choices available?
10. Are the fruit juices 100% juice?
11. Are foods (especially desserts) high in fat, sugar, and/or sodium balanced with choices lower in these nutrients?
12. Is a soft margarine available that does not contain trans fat?
13. Are unsweetened beverages available?

Look at the Mediterranean, Asian, and Latin American Diet Pyramids (Figs. 2-18 to 2-20).

4. **Read food labels**

   The Food and Drug Administration regulates labels on all packaged foods except for meat, poultry, and egg products (these are under the USDA).

   All food labels contain at least

   The name of the food
   A list of ingredients in descending order by weight
   The net contents or net weight—this is the quantity of the food itself without the packaging
   The name and place of business of the manufacturer, packer, or distributor
   Nutrition information

   Food labels must use the common names of the 8 most common allergenic foods: milk, eggs, fish, shellfish, tree nuts, peanuts, wheat, and soybeans.

   Read the "**Nutrition Facts**" on a food label (Fig 2-22). **Daily Value** is a set of nutrient-intake values used as a reference for expressing nutrient content on nutrition labels. Foods that contain 5% or less of the Daily Value for a nutrient are generally considered low in that nutrient. Foods that contain 20% or more of the Daily Value for a nutrient are generally considered high in that nutrient.

   Nutrient content claims, such as "good source of calcium," are claims on labels about the nutrient composition of a food. They are regulated by the Food and Drug Administration and must follow legal definitions as seen in Figure 2-24.

   Health claims state that certain foods or components of foods may reduce the risk of a disease or health-related condition. Health claims are ranked as follows: A (significant scientific agreement), B (moderate, evidence is not conclusive), C (low, evidence is limited and not conclusive), and D (extremely low, little scientific evidence supporting this claim). Qualified health claims (always ranked B, C, or D) require a disclaimer or other qualifying language to ensure that they do not mislead consumers.

   You may see claims on labels that refer to a broad class of foods and a disease. For example, "diets rich in fruits and vegetables may reduce the risk of some types of cancer." This is a statement using current dietary guidance. Truthful, non-misleading dietary guidance statements may be used on food labels, and do not have to be reviewed by the FDA. However, once the food is marketed with the statement, FDA can consider whether the statement meets the requirement to be truthful and not misleading.

5. **Compare portion sizes on food labels with MyPyramid**

   Portion sizes in the U.S. tend to be large.
   Portion sizes on food labels are not always the same as on MyPyramid. MyPyramid is purposely designed with a few serving sizes for each group to make it easy to use. Food label serving sizes make it possible to compare different brands.

6. **Food Facts: Nutrient Analysis of Recipes**
   Computer software is a wonderfully quick way to analyze the amount of nutrients in a recipe or food.

7. **Hot Topic: Quack! Quack!**
   Food quackery can be defined as "the promotion for profit of special foods, products, processes, or appliances with false or misleading health or therapeutic claims."

   Review how to recognize a quack.
   1. *Their promises are too good to be true.*
   2. *They use dubious diagnostic tests, such as hair analysis, to detect supposed nutritional deficiencies and illnesses.*
   3. *They rely on testimonials as proof of effectiveness.*
   4. *Some use food essentially as medicine.*
   5. *They often lack any valid medical or health care credentials.*
   6. *They come across more as salespeople than as medical professionals.*
   7. *They offer simple answers to complex problems.*
   8. *They claim to be persecuted and sabotaged by governmental and medical institutions.*
   9. *They make dramatic statements that are refuted by reputable scientific organizations.*
   10. *Their theories and promises are not written in medical journals using a peer review process, but appear in books written only for the lay public.*

   Registered Dietitians represent the largest and most visible group of professionals in the nutrition field. RDs are recognized by the medical profession as legitimate providers of nutrition care.

NUTRITION WEB EXPLORER

MyPyramid: www.mypyramid.gov
At the MyPyramid website, perform each of the following:

1. Under "MyPyramid Plan," enter your age, gender, and activity level to see how many total kcalories, servings from each food group, and discretionary kcalories you are allowed.
2. Next, click on "Meal Tracking Worksheet" to make a copy of your MyPyramid worksheet. Write down everything you eat in one day, and compare it to the recommendations.
3. To get an idea of portion sizes, click on "Inside the Pyramid," and then on a food group such as Meats and Beans. Then click on "What's in the Meat and Beans Group?" Next click on, "View Meat and Beans Food Gallery." Do this for each food group.
4. To get a nutrient analysis of your one day intake, click on "MyPyramid Tracker," and register as a new user. Then click on "Assess Your Food Intake." Enter in the foods and serving sizes that you ate yesterday. Once you have done that, click on "Save and Analyze" and obtain each of these reports: Meeting 2005 Dietary Guidelines, Nutrient Intake, and MyPyramid Recommendation.
5. You can also plan menus with the MyPyramid Menu Planner. The Menu Planner shows whether your food choices are balanced for the day, or on average over a week. You can also use it to help plan upcoming meals to meet MyPyramid goals.

American Dietetic Association: www.eatright.org
Visit the ADA website and get its "Tip of the Day." Also use "Find a Nutrition Professional" to find a list of dietitians in your area. Write below the Tip of the Day and a RD in your area.

Dietitians of Canada: www.dietitians.ca/
Visit the Dietitians of Canada website, and click on "Find a Nutrition Professional" to find a list of dietitians in your area. On the home page, click on "Eat Well, Live Well," then click on "Let's Make a Meal." This interactive program let you choose various menu items for breakfast, lunch, dinner, and snacks and compares your choices with the Canadian food guide. How does your meal compare?

Quackwatch: www.quackwatch.com
Visit this website and click on "25 Ways to Spot It" under "Quackery." What are 10 ways to spot quackery?

International Dietary Guidelines:
http://fnic.nal.usda.gov/nal_display/index.php?info_center=4&tax_level=2&tax_subject=270&topic_id=1339

Visit this government website about the Dietary Guidelines for Americans. On the home page, under "International Dietary Guidance," click on "Dietary Guidelines from Around the World" (at the bottom of the page) and read about the diets of people in another country. Describe the dietary guidelines from another country.

Food Label Quiz: www.cfsan.fda.gov/~dms/flquiz1.html
Take the "Test Your Food Label Knowledge Quiz" at this government website. How did you score?

University of Florida Libraries Tips on Web Search:
http://web.uflib.ufl.edu/admin/wwwtips.pdf
This article presents eight questions to keep in mind when searching for reliable information on the Web. List at least 5 questions below.

CHAPTER REVIEW QUIZ

Key Terms: Matching

1. Dietary recommendations
2. Food guides
3. Dietary Guidelines for Americans
4. Discretionary kcalories
5. Daily value
6. Nutrient content claims
7. Health claims
8. Qualified health claims

a. Claims on food labels about the nutrient composition of a food, regulated by the Food and Drug Administration
b. Guidelines that tell us the kinds and amounts of foods that constitute a nutritionally adequate diet
c. A set of nutrient-intake values developed by the Food and Drug Administration that are used as a reference for expressing nutrient content on nutrition labels
d. Guidelines that discuss specific foods and food groups to eat for optimal health
e. A set of dietary recommendations for Americans that is periodically revised.
f. Health claims graded B, C, or D that require a disclaimer or other qualifying language to ensure that they do not mislead consumers
g. Claims on food labels that state certain foods or food substances-as part of an overall healthy diet-may reduce the risk of certain diseases
h. The balance of kcalories you have after meeting the recommended nutrient intakes by eating foods in low-fat or no added sugar forms

Multiple Choice

1. According to the 2,000 Dietary Guidelines, in order to lose weight, one must pay attention to:
    a. Kcalories
    b. Portions of fat, carbohydrates, and protein in the diet
    c. Kcalories and portions of fat in the diet
    d. Both a and b

2. According to the 2000 Dietary Guidelines, it is recommended that one consume:
    a. Less than 1,500 mg of sodium per day
    b. Less than 2,000 mg of sodium per day
    c. Less than 2,300 mg of sodium per day
    d. Less than 2,500 mg of sodium per day

3. According to the 2000 Dietary Guidelines, it is recommended that total fat intake in one's diet:
    a. Should not exceed 15% of total kcalories
    b. Should be between 10 to 25% of total kcalories
    c. Should be between 20 to 35% of total kcalories
    d. There is no recommendation

4. According to the 2000 Dietary Guidelines, in order to reduce the risk of chronic disease in adulthood, it is recommended that a person exercise:
    a. At least 30 minutes 3 days per week
    b. At least 30 minutes most days
    c. At least 60 minutes most days
    d. At least 60–90 minutes most days

5. According to the 2000 Dietary Guidelines, nutrients that may be deficient for children and adolescents include:
    a. Calcium and phosphorus
    b. Beta-carotene and vitamin K
    c. Calcium and beta-carotene
    d. Calcium and magnesium

True/False

1. According to the 2000 Dietary Guidelines, the Acceptable Macronutrient Distribution Range for carbohydrates is 55 to 65 percent. 45-65
    a. True   b. False

2. On average, the higher your salt intake, the higher your blood pressure.
    a. True   b. False

3. According to the 2000 Dietary Guidelines, alcohol may have beneficial effects when consumed in moderation.
   a. True b. False

4. According to the 2000 Dietary Guidelines, decreased potassium intake can prevent or delay the onset of high blood pressure.
   a. True b. False

5. In order to symbolize a personalized approach to healthy eating and exercise, there are 12 MyPyramids ranging from 1,000 to 3,200 kcalories.
   a. True b. False

6. Thiamin, riboflavin, and niacin play a key role in metabolism, and are essential for a healthy nervous system.
   a. True b. False

7. Foods labeled with the words "multigrain," "stone ground,", "bran,", or "100% wheat" are usually not whole-grain products.
   a. True b. False

8. Vitamin E helps heal cuts and wounds and keep teeth and gums healthy.
   a. True b. False

9. Vitamin C helps the body absorb iron.
   a. True b. False

10. Most fruits are naturally in kcalories even though they contain natural sugars.
    a. True b. False

Short Answer

1. Why is the consumption of whole fruits recommended over fruit juice for the majority of total daily intake?

2. How much milk, yogurt, and cheese constitute one serving of milk?

3. Which vitamin functions in the body to maintain proper levels of calcium and phosphorus in the blood, thereby helping to build and maintain bones?

4. Which cholesterol is considered "bad" cholesterol? What will result from high levels of this cholesterol?

5. What does MyPyramid consider most solid fats and all added sugars?

6. Which types of fats should one consume primarily in the diet?

7. Which types of fat increase the risk of heart disease?

8. Which three types of activities are beneficial to one's health and can decrease the risk of chronic diseases such as diabetes mellitus and hypertension?

9. Which agency regulates labels on all packaged foods except for meat, poultry, and egg products?

10. Which eight ingredients account for 90 percent of all food allergies?

STUDENT WORKSHEET 2-1

TOPIC: Introduction to MyPyramid

1. Go to www.mypyramid.gov.

2. Click on "Animation," and then click on "Long Version with Audio." Watch and listen to the animated feature that describes the basics of MyPyramid.

3. Write down below the food group represented by each of the following colors.

   a. Orange: ______________________________

   b. Green: ______________________________

   c. Red: ______________________________

   d. Blue: ______________________________

   e. Purple: ______________________________

4. Next, click on "Inside the Pyramid" on the home page. Click on each part of the Pyramid and write down below at least 2 points listed on the webpage for each of the food groups. For example:

   Grains:
   1. Make half your grains whole.
   2. Eat at least 3 ounces of whole grain bread, cereal, crackers, rice, and pasta every day.

   Food Group: ______________________________________

   1.

   2.

   Food Group: ______________________________________

   1.

   2.

Food Group: ____

1.

2.

Food Group: ____

1.

2.

5. On the home page, look at the box entitled "MyPyramid Plan." Fill in your age, then select your sex and physical activity level, and also fill in height and weight. Click "Submit." Print out a copy of your results by clicking on: "Click here to view and print a PDF version of your results."
How many kcalories are your allowed? How much from each food group are you allowed each day?

STUDENT WORKSHEET 2-2

TOPIC: MyPyramid Food Groups

1. Go to www.mypyramid.gov.

2. Click on "Inside the Pyramid." Next, click on "Grains" to fill in the information needed below. For each "Sample Food" in the first column, write down its serving size in the second column (click on "What Counts as an Ounce or Cup?"). For each food group, be sure to click on, for example, "View Grains Food Gallery" on the food group's home page. This feature gives photos of actual serving sizes. Click on "Health Benefits and Nutrients" for the right-hand columns.

GRAINS

Sample Foods	What counts as an ounce?	Health Benefits	Nutrients in Grains

3. Do the same for Vegetables, Fruits, Milk, and Meat & Beans.

VEGETABLES

Sample Foods	What counts as 1 cup?	Health Benefits	Nutrients in Vegetables

FRUITS

Sample Foods	What counts as 1 cup?	Health Benefits	Nutrients in Fruits

MILK

Sample Foods	What counts as 1 cup?	Health Benefits	Nutrients in Milk

MEATS & BEANS

Sample Foods	What counts as 1 ounce?	Health Benefits	Nutrients in Meats & Beans

STUDENT WORKSHEET 2-3

TOPIC: MyPyramid Tips

1. Go to www.mypyramid.gov.

2. Click on "Tips and Resources." Fill in 2 tips you find useful for each of the following topics.

Make half your grains whole	
Vary your veggies	
Focus on fruit	
Get your calcium rich foods	
Go lean with protein	
Find your balance between food and physical activity	

STUDENT WORKSHEET 2-4

TOPIC: Personal Diet Analysis

Your instructor will tell you whether to complete #1 or #2 below. Both exercises require that you have a list of all the foods and beverages (including serving size) you consumed in one day.

1. MyPyramid Worksheet
   - A. Go to www.MyPyramid.gov.
   - B. Look at the box entitled "MyPyramid Plan." Fill in your age, then select your sex and physical activity level. Click "Submit."
   - C. Click on "Meal Tracking Worksheet" on the right-hand side, and follow the directions on it. This page shows how many servings from each food group you should be eating each day. If you fill in what you ate yesterday or today, you can compare that to the goals based on your kcalorie pattern.

2. MyPyramid Tracker
   - A. Go to www.MyPyramid.gov. Look at the box entitled "MyPyramid Plan." Fill in your age, then select your sex and physical activity level. Click "Submit."
   - B. This page now shows you how many kcalories and how much of each food group is appropriate each day. On the right-hand side of this page, click on "MyPyramid Tracker."
   - C. Before you can use MyPyramid Tracker, you need to register. At the bottom of the page, click on "New User Registration" and go through the process. Once you have answered all the questions, click on "Proceed to Food Intake."
   - D. Now you can start entering food and beverages. Enter your first food, such as banana, in the box and click on "Search." The program will show you a variety of food items—pick the one that is closest to what you ate. In the case of bananas, click on "Bananas, fresh" and the food will be put on the right hand side of the page. Next, click on "Select Quantity" and select a serving size and enter in the number of servings. Click on "Enter Foods" to return to the page where you will add more foods to your list.
   - E. Once you have entered in your food list, click on "Save and Analyze." You can now print out these 3 reports:
     1. Meeting 2005 Dietary Guidelines
     2. Nutrient Intakes
     3. MyPyramid Recommendations
   - F. Summarize the findings below about your one-day diet. Compare your kcalorie intake to your MyPyramid plan. Which nutrients were high? Which nutrients were low? Which food groups are high or low? Which foods could you add or subtract to your diet to improve the overall quality?

STUDENT WORKSHEET 2-5

TOPIC: MyPyramid Sleuth

Use the www.MyPyramid.gov website to answer the following questions.

1. How are oils different from solid fats?

2. Why is it importance to consume oils?

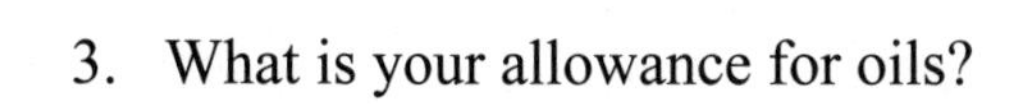

3. What is your allowance for oils?

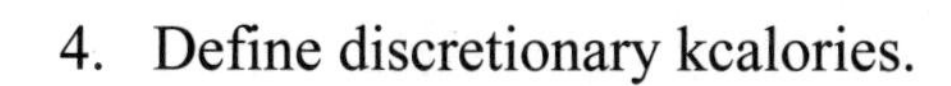

4. Define discretionary kcalories.

5. What is your discretionary kcalorie allowance?

6. What nutrients are important for vegetarians?

7. How do you count 1 slice of thin-crust pizza and 1 peanut butter and jelly sandwich using MyPyramid?

8. How much physical activity should you be getting every day?

STUDENT WORKSHEET 2-6

TOPIC: Food Labels

Use the label below to answer the following questions.

Nutrition Facts

Serving Size 4 cookies (31g)
Servings Per Container about 9

Amount Per Serving	
Calories 160	Calories from Fat 80
	% Daily Value*
Total Fat 9g	**13%**
Saturated Fat 6g	**28%**
Cholesterol 0mg	**0%**
Sodium 140mg	**6%**
Total Carbohydrate 20g	**7%**
Dietary Fiber 1g	**5%**
Sugars 11g	
Protein 1g	
Vitamin A 0% •	Vitamin C 0%
Calcium 0% •	Iron 2%

* Percent Daily Values are based on a 2,000 calorie diet. Your daily values may be higher or lower depending on your calorie needs:

	Calories:	2,000	2,500
Total Fat	Less than	65g	80g
Sat Fat	Less than	20g	25g
Cholesterol	Less than	300mg	300mg
Sodium	Less than	2,400mg	2,400mg
Total Carbohydrate		300g	375g
Dietary Fiber		25g	30g

Calories per gram:
Fat 9 • Carbohydrate 4 • Protein 4

INGREDIENTS: ENRICHED FLOUR (WHEAT FLOUR, NIACIN, REDUCED IRON, THIAMINE MONONITRATE, RIBOFLAVIN), SUGAR, VEGETABLE SHORTENING (CONTAINS ONE OR MORE OF THE FOLLOWING PARTIALLY HYDROGENATED OILS: PALM KERNEL, SOYBEAN, COTTONSEED), COCOA (PROCESSED WITH ALKALI), CARAMEL COLOR, LEAVENING (SODIUM BICARBONATE, MONOCALCIUM PHOSPHATE, AMMONIUM BICARBONATE), HIGH FRUCTOSE CORN SYRUP, SALT, WHEY, SOY LECITHIN (EMULSIFIER), PEPPERMINT OIL, NATURAL AND ARTIFICIAL FLAVOR

1. How can you prove that the kcalories from fat is actually 80 as on the label?

2. Name 1 nutrient that this food is high in and 1 nutrient that it is low in by using the % Daily Value.

3. The % Daily Value is based on a diet of how many Kcalories? Where did you find this information?

4. How many grams of fat are in 2 cookies?

5. Is 1 serving of cookies high in sodium or cholesterol?

6. About how many cookies are in the entire box?

7. What is the Daily Value for dietary fiber for someone who can eat 2,500 kcal/day?

CHAPTER 3 CARBOHYDRATES

LEARNING OBJECTIVES

Upon completion of the chapter, the student should be able to:

1. Identify the functions of carbohydrates
2. List important monosaccharides and disaccharides and give examples of foods in which each is found
3. Identify foods high in natural sugars, added sugars, and fiber
4. List the potential health risks of consuming too much added sugar
5. Identify food sources of starch and list the uses of starch in cooking
6. Distinguish between the two types of dietary fiber and list examples of food containing each one
7. Describe the health benefits of a high-fiber diet
8. Describe how carbohydrates are digested, absorbed, and metabolized by the body
9. State the dietary recommendations for carbohydrates
10. Identify foods as being made from whole grains or refined grains
11. Discuss the nutritional value and use of grains and legumes on a menu
12. Recognize alternatives to sugar in foods

CHAPTER OUTLINE

1. **Carbohydrate basics**

   Carbohydrates are the major components of most plants as plants make carbohydrates from the carbon dioxide in the air and water from the soil in a process known as photosynthesis.

   There are two categories: simple carbohydrates are sugars—including both natural and refined sugars, and complex carbohydrates (or polysaccharides) include starch and fiber, long chains of sugars.

2. **Functions of carbohydrates**

   Carbohydrate is the primary source of energy for your body. The central nervous system relies almost exclusively on glucose and other simple carbohydrates for energy. Some glucose is stored in the liver and muscles as glycogen.

   Carbohydrate is also important to help the body use fat efficiently. When fat is burned for energy without any carbohydrate present, the process is incomplete and could result in ketosis (excessive level of ketone bodies in the blood and urine). You need at least 100 to 150 grams of carbohydrates daily to prevent protein and fat from being burned for fuel and to provide glucose to the central nervous system and red blood cells. Carbohydrates are protein-sparing.

   Carbohydrates are part of various materials found in the body such as connective tissues, some hormones and enzymes, and genetic material.

   Fiber promotes the normal functioning of the intestinal tract, lowers blood cholesterol, and is associated with a reduced risk of developing type 2 diabetes.

3. **Monosaccharides and disaccharides**
   Monosaccharides are the building blocks of other carbohydrates and include the simple sugars glucose, fructose, and galactose.

   Glucose (dextrose)—body's main fuel source, most carbohydrates are converted to glucose which the body distributes to the cells as blood glucose, found in grapes, in honey and in trace amounts in most plant foods

   Fructose—sweetest natural sugar, found in fruits and honey

   Galactose—does not occur alone in nature, linked to glucose in milk sugar, a disaccharide

   Most naturally occurring carbohydrates contain 2 or more monosaccharide units linked together. Disaccharides, the double sugars, include sucrose, maltose, and lactose.

   Sucrose (cane sugar, table sugar, granulated sugar)—simply glucose and fructose linked together, table sugar is 99% pure sugar and provides virtually no nutrients, occurs naturally in some fruits and vegetables, table sugar is made from sugarcane or sugar beets and provides virtually no nutrients for its 16 kcalories per teaspoon

   Maltose—2 glucose units bonded together, does not occur in nature to an appreciable extent

   Lactose (milk sugar)—found naturally only in milk, lactose is not very sweet

4. **Added sugars**
   Added sugars include white sugar, high-fructose corn syrup, and other sweeteners added to foods in processing, as well as sugars added to foods at the table.

   High-fructose corn syrup has been treated with an enzyme that converts part of the glucose to fructose. While table sugar consumption has dropped; high-fructose corn syrup consumption has increased and is used to sweeten regular soft drinks and is frequently used in fruit drinks, sweetened teas, cookies, jams and jellies, and syrups.

   Major sources of added sweeteners in the diet are soft drinks, candy and sugars, baked goods, fruit drinks, and dairy desserts and sweetened milk.

   To find out whether a food contains added sugar and how much, look at the ingredient list. The number of grams of sugar on the label includes naturally occurring sugars and added sugars.

   Besides sweetening, added sugars prevent spoilage in jams and jellies, help baked goods brown and retain moisture, and provide food for yeast.

5. **Added sugars and health issues**
   Dental caries—Sugar contributes to dental caries. The more often sugars and starches are eaten, the more often bacteria ferment these carbohydrates and produce acid which eats away at teeth.

   Food such as dried fruits, breads, cereals, cookies, crackers, and potato chips increase the chances of dental caries when eaten frequently. Foods that do not seem to cause cavities include cheese, peanuts, sugar-free gum, some vegetables, meats, and fish. To prevent dental caries, brush your teeth twice a day, floss your teeth every day, try to limit sweets to mealtime, and see your dentist regularly.

   Obesity—Although there is no research stating that added sugars cause obesity, added sugars are undoubtedly a factor in rising obesity rates among adults and children. Individuals who consume food or beverages high in added sugars tend to consume more kcalories than those who consume low amounts of added sugars, and also tend to consume lower amounts of vitamins and minerals. Just add one 12-ounce can of soft drinks to your diet every day for a year, and you will gain 15 pounds! Empty kcalorie foods provide few nutrients for the number of kcalories they provide.

   Diabetes—There is no evidence that total sugar intake is associated with the development of diabetes.

   Heart Disease—A moderate intake of sugars does not increase heart disease risk, but diets high in fructose and sucrose seem to increase blood levels of fat and cholesterol, which then increase the risk of heart disease.

   Hypoglycemia—It occurs most often in people who have diabetes and take insulin. Diet includes regular, balanced meals with moderate amounts of refined sugars and sweets.

   Hyperactivity in children—Research has failed to show that high sugar intake increases hyperactivity.

6. **Lactose intolerance**
   Lactase deficiency results in abdominal cramps, bloating, and diarrhea, that normally occur within 30 minutes to 2 hours. The symptoms are normally cleared up within 2 to 5 hours.

   Lactose intolerance seems to be an inherited problem among Asian Americans, Native Americans, African Americans, and Latinos.

   There is tremendous individual variation as far as what foods can be tolerated and when.

   Lactose-reduced milk, yogurt, and hard cheeses are usually well-tolerated.

7. **Complex carbohydrates: starches**
   Starch is made of many chains of hundreds to thousands of linked glucoses. It is found only in plant foods such as cereal grains and foods made from them, root vegetables, and dried beans and peas.

Starchy foods must generally be cooked to make them better tasting and digestible. They are used as thickeners in cooking because starch undergoes gelatinization (process in which starches, when heated in liquid, absorb water and swell in size).

Starch contributes to tooth decay.

8. **Complex carbohydrates: fiber**
The Food and Nutrition Board defines dietary fiber as the polysaccharides found in plant foods that are not digested and absorbed. Their definition also includes lignin, a part of plant cells that is not technically a polysaccharide.

Like starch, most fibers are chains of bonded glucose units, but what's different is that the units are linked with a chemical bond that our digestive enzymes can't break down. Our digestive enzymes can't break down the glucose units in fiber, except that some fiber is digested by bacteria in the large intestine.

Fiber is found only in plant foods; it does not appear in animal foods. Legumes and whole grains are excellent sources, as are fruits and vegetables and nuts and seeds.

Most foods contain both soluble and insoluble fiber

Type of Fiber	Food Sources	Health Benefits
Soluble (Viscous)	Beans and peas Some cereal grains such as barley, oats, rye Many fruits such as citrus fruits, pears, apples, grapes Many vegetables such as Brussels sprouts and carrots	• Traps carbohydrates to slow digestion and absorption of glucose • Binds to cholesterol in gastrointestinal tract • Reduces risk of diabetes and heart disease
Insoluble (Nonviscous)	Wheat bran Whole grains, such as whole wheat and brown rice Many vegetables Many fruits Beans and peas Seeds	• Increases fecal weight, so feces travels quickly through the colon • Provides feeling of fullness • Helpful to prevent and treat constipation, diverticulosis, and hemorrhoids • Helpful in weight management • May help reduce risk of heart disease

Diets high in fiber and low in saturated fat and cholesterol are associated with a reduced risk of diabetes, heart disease, digestive disorders, and certain cancers. When adding fiber to the diet, do so gradually to give the intestinal tract time to adapt and drink lots of fluid.

9. **Nutrition Science Focus: Carbohydrates**
   Carbohydrates you eat are absorbed mostly as glucose into the blood stream. The level of glucose is kept within a certain range so that your body's cells will have adequate supply of this important fuel. As you are absorbing nutrient from food, insulin is the dominant hormones. Insulin helps glucose enter the cells and also stimulates certain cells to make glycogen or fat. When blood glucose levels fall between meals and at night, insulin secretion slows and glucagon becomes more dominant. Glucagon stimulates the liver to convert glycogen to glucose and release the glucose into the bloodstream. In diabetes, the pancreases either produces too little or no insulin.

   Soluble fiber (oats, legumes, fruits, and vegetables) lowers blood cholesterol and traps carbohydrates to slow digestion and absorption of glucose, thereby helping blood glucose levels stay constant. Insoluble fiber (wheat bran, whole grains, fruits, and vegetables) is important for gastrointestinal health and it helps prevent and treat constipation, diverticulosis, and hemorrhoids.

10. **Digestion, absorption, and metabolism of carbohydrates**
    During digestion, various enzymes break down starch and sugars to monosaccharides, which are then absorbed across the intestinal wall. In the liver, fructose and galactose will be converted to glucose or further metabolized to make glycogen or fat.

    Although human enzymes can't digest most fibers, some bacteria in the large intestine can digest soluble fibers. As they digest the soluble fibers, the bacteria produce gas and small fat particles that are absorbed. Soluble fiber slows the digestive process and slows the absorption of glucose into the blood.

11. **Dietary recommendations for carbohydrates**
    The RDA for carbohydrate is 130 grams/day for adults and children over 1 year of age. It is based on the minimum amount of carbohydrates needed to supply the brain with enough glucose. Added sugars should not exceed 25% of total kcal.

    The AMDR for carbohydrate is 45 to 65% of kcalories from carbohydrates for adults and children over 1 year of age.

    The Dietary Reference Intakes recommend that added sugars not exceed 25% of total kcalories. The World Health Organization suggests limiting added sugars to 10%.

    The AI for total fiber for men and women (19 to 50 years old) is set at 38 and 25 grams/day respectively. The AI for total fiber is based on 14 grams/1,000 kcalories.

    The Dietary Guidelines for Americans recommend 3 or more servings daily of whole grains.

12. **Ingredient Focus: High-Fiber Grains and Legumes**

Grains are the edible seeds of cultivated grasses such as wheat, corn, rice, rye, barley, and oats. All grains have a large center area high in starch known as the endosperm. At one end of the endosperm is the germ, the area of the kernel that sprouts when allowed to germinate. The germ is rich in vitamins and minerals and contains some oil. The bran, containing much fiber and other nutrients, covers both the endosperm and the germ.

Whole grains contain the endosperm, germ, and bran. Examples of whole grains to look for on food labels:

- brown rice
- bulgur (cracked wheat)
- oatmeal
- whole oats
- whole wheat
- whole rye
- whole hulled barley
- whole grain corn
- popcorn

Refined grains are grains in which the bran and germ are separated from the endosperm. White flour is made only from the endosperm of the wheat kernel. Whole-whole flour is made from the whole grain. When you compare the nutrients in whole grains and refined grains, whole grains are always a far more nutritious choice.

Grains, such as rice, are low or moderate in kcalories, high in starch and fiber (if whole grain), low in fat, moderate in protein, and full of vitamins and minerals. They are also inexpensive and can be quite profitable. There are many types of rice available, and an endless number of dishes they can be used in.

Legumes include all sorts of dried beans, peas, and lentils. Legumes are high in complex carbohydrates and fiber, low in fat and sodium, cholesterol free, and a good source of vitamins and minerals. Many varieties of legumes are available.

Grains and legumes are all cooked in water.

13. **Culinary Science**

Starchy foods are used extensively as thickeners in cooking because starch undergoes a process called gelatinization when heated in liquid. When starches gelatinize, the granules absorb water and swell, making the liquid thicken. Around the boiling point, the granules have absorbed a lot of water and burst, letting starch out into the liquid. When this occurs, the liquid quickly becomes still thicker. Common starchy thickeners include flour, cornstarch, and arrowroot.

14. **Food Facts: Glycemic Index**

Glycemic response refers to how quickly and how high your blood sugar rises after eating. Any number of factors influence how high your blood glucose will rise, such as the amount of carbohydrate eaten, the type of sugar or starch, and the presence of fat and other substances that slow down digestion. A low glycemic response (meaning your blood sugar rises slowly and modestly) is preferable to a high glycemic response. Eating lots of foods with a low glycemic response is important for people with diabetes and seems to decrease the risk of heart disease, type 2 diabetes, as well as enhancing weight management.

The concept of glycemic index was created to identify how selected foods affect your blood glucose level. Glucose is assigned a value of 100. The higher the glycemic index is, the more the food increases your blood glucose level. A shortcoming of the glycemic index is that we usually eat more than one food at a time.

15. **Hot Topic: Alternatives to Sugar: Artificial Sweeteners and Sugar Replacers**
Artificial sweeteners include saccharin, aspartame, acesulfame-K, sucralose, and Neotame. Neotame is about 7,000 to 13,000 times sweeter than sugar. It is a white crystalline powder that is heat stable and can be used as a tabletop sweetener as well as in cooking.

The artificial sweeteners alitame and cyclamate are awaiting approval from the FDA. *Stevioside* is a naturally sweet extract from the leaves of the stevia bush found in South America. The extract is 300 times sweeter than sucrose. The FDA has now approved stevia as a Generally Regarded As Safe (GRAS) additive, so you will find it at the supermarket and in restaurants as a tabletop sweetener.

Sugar replacers, also called polyols or sugar alcohols, are a group of carbohydrates that are sweet and occur naturally in plants. Many sugar replacers, such as xylitol, have been used for years in products such as sugar-free gums, candy and fruit spreads. Sugar replacers don't provide as many kcalories as sugar—usually about 2/gram compared with 4/gram from sugar. They don't promote tooth decay and cause smaller increases in blood glucose and insulin levels than sugar. Some sugar replacers, such as sorbitol and mannitol, may have a laxative effect and there is a warning on the label.

NUTRITION WEB EXPLORER

Joslin Diabetes Center: www.joslin.org/Beginners_guide_2854.asp
Joslin Diabetes Center is an excellent site to learn almost anything about diabetes. Read "Carbohydrate Counting 101" and then complete the exercise "Meal Planning." What did you learn?

National Institutes of Health on Hypoglycemia:
www.nlm.nih.gov/medlineplus/hypoglycemia.html
On this government website, click on "Hypoglycemia-Interactive Tutorial" to learn more about this topic. Write a paragraph below on what you learned.

National Institutes of Health on Lactose Intolerance:
http://digestive.niddk.nih.gov/ddiseases/pubs/lactoseintolerance/
Use this informative site to learn more about the causes and treatment of lactose intolerance. Find out how many grams of lactose are in reduced-fat milk and compare that to yogurt.

Whole Grains Council:
http://wholegrainscouncil.org/whole-grains-101/whole-grains-a-to-z
Read about a huge variety of whole grains. Write up a description of three whole grains that are new to you.

CHAPTER REVIEW QUIZ

Key Terms: Matching

1. Carbohydrate
2. Glucose
3. Glycogen
4. Ketone bodies
5. Ketosis
6. Fructose
7. Galactose
8. Maltose
9. Lactose
10. Diabetes
11. Insulin
12. Glycemic response
13. Glycemic index
14. Lactase
15. Gelatinization
16. Soluble fiber
17. Insoluble fiber
18. Endosperm
19. Germ
20. Bran
21. Phytochemical

a. Classification quantifying the blood glucose response after eating carbohydrate-containing foods
b. A hormone that increases the movement of glucose from the bloodstream into the body's cells
c. A group of organic compounds that cause the blood to become too acidic as a result of fat being burned for energy without any carbohydrates present
d. The most significant monosaccharide; the body's primary source of energy
e. A classification of fiber that includes cellulose, lignin, and resistant starch; they generally form the structural parts of plants
f. A large class of nutrients, including sugars, starch, and fibers, that function as the body's primary source of energy
g. A disaccharide found in milk and milk products that is made of glucose and galactose
h. The storage form of glucose in the body; it is stored in the liver and muscles
i. In cereal grains, the area of the kernel rich in vitamins and minerals that sprouts
j. Minute substances in plants that may reduce the risk of cancer and heart disease when eaten often
k. A classification of fiber that includes gums, mucilages, pectin, and some hemicelluloses; they are generally found around and inside plant cells
l. A monosaccharide found linked to glucose to form lactose, or milk sugar
m. An enzyme needed to split lactose into its components in the intestines
n. A process in which starches, when heated in liquid absorb water and swell in size
o. A monosaccharide found in fruits and honey
p. In cereal grains, the part that covers the grain and contains much fiber and other nutrients
q. Two glucose units bonded together
r. A disorder in which the body does not metabolize carbohydrate properly due to inadequate or ineffective insulin
s. In cereal grains, a large center area high in starch
t. How quickly and how high blood glucose rises after eating
u. Excessive level of ketone bodies in the blood and urine

Multiple Choice

1. ________________, also known as dextrose, is the most abundant sugar found in nature.
    a. Sucrose
    b. Glucose
    c. Fructose
    d. Maltose

2. ________________ is a single sugar that links with glucose to make milk sugar.
    a. Lactose
    b. Glucose
    c. Galactose
    d. Dextrose

3. ________________ spares protein from being burned for energy.
    a. Galactose
    b. Fructose
    c. Fiber
    d. a and b

4. ________________ consists of two bonded glucose units and is produced in the manufacture of beer.
    a. Maltose
    b. High fructose corn syrup
    c. Dextrose
    d. Sucrose

5. ________________ is only 75% as sweet as sucrose and, due to low cost, is used many baked goods and other foods.
    a. Table sugar
    b. Sugar cane
    c. Brown sugar
    d. High fructose corn syrup

True/False

1. Foods that do not seem to cause cavities include cheese, peanuts, and fish.
    a. True   b. False

2. There is no evidence that total sugar intake is associated with the development of diabetes mellitus.
    a. True   b. False

3. Any amount of refined sugars has the ability to cause swings in blood glucose levels.
   a. True   (b.) False

4. Most people produce less lactose after the age of two, although many will not experience symptoms until they are much older.
   a. True   (b.) False

5. Symptoms of hypoglycemia include anxiety and stress symptoms.
   (a.) True   b. False

6. Plants such as peas store glucose in the form of glycogen.
   a. True   (b.) False

7. Starch from whole-grain sources is preferable to starch found in refined grains such as white flour.
   (a.) True   b. False

8. The amount of fiber in a plant varies among plants, and may vary within species, depending on the plant's maturity.
   (a.) True   b. False

9. Hemorrhoids are a condition in which small pouches form in the colon wall.
   a. True   (b.) False

10. Phytochemicals are substances in plants that reduce the risk of cancer and heart disease when eaten often.
   (a.) True   b. False

Short Answer

1. What is the sweetest natural sugar?

   fructose

2. Which sugars are considered simple carbohydrates?

   monosaccharides + disaccharides

3. What amount of carbohydrates must be eaten daily to prevent protein and fat from being burned for fuel?

   100 - 150 grams/day

4. During which process do plants convert energy from sunlight into energy stored in carbohydrate?

   photosynthosis

5. Which foods are the major sources of added sugars in the American diet?

Soft drinks, candies & sugars, baked goods, fruit drinks and dairy desserts

6. Which food contains more sugar, kidney beans or peanuts?

Kidney beans

7. What, during digestion, breaks down starch and sugars into monosaccharides?

enzymes

8. What is one way to classify fibers?

One can classify fiber by wether it is soluble in water (soluble + insoluble)

9. How many whole grains are recommended daily, according to MyPyramid?

3 servings

10. Which foods are categorized as legumes?

dried beans, peas, lentils

STUDENT WORKSHEET 3-1

TOPIC: Reading Food Labels from Breakfast Cereals

Use the following "Nutrition Facts" and ingredients lists to answer questions 1–8.

Nutrition Facts
Serving Size 1 cup (30g)
Servings Per Container About 19

Amount Per Serving		with ½ cup skim milk
Calories	120	160
Calories from Fat	15	15
	% Daily Value**	
Total Fat 1.5g*	2%	3%
Saturated Fat 0g	0%	0%
Polyunsaturated Fat 0.5g		
Monounsaturated Fat 0.5g		
Cholesterol 0mg	0%	1%
Sodium 270mg	11%	14%
Potassium 90mg	3%	8%
Total Carbohydrate 24g	8%	10%
Dietary Fiber 2g	8%	8%
Soluble Fiber less than 1g		
Sugars 11g		
Other Carbohydrate 11g		
Protein 3g		

CEREAL #1
Ingredients: Whole grain oats, sugar, oat bran, modified corn starch, honey, brown sugar syrup, salt, ground almonds, calcium carbonate, trisodium phosphate, wheat flour, vitamins and minerals.

Nutrition Facts
Serving Size ¾ cup (30g)
Servings Per Container About 17

Amount Per Serving		with ½ cup skim milk
Calories	120	160
Calories from Fat	10	10
	% Daily Value**	
Total Fat 1g*	2%	2%
Saturated Fat 0g	0%	0%
Polyunsaturated Fat 0g		
Monounsaturated Fat 0g		
Cholesterol 0mg	0%	1%
Sodium 270mg	11%	14%
Potassium 55mg	2%	7%
Total Carbohydrate 25g	8%	10%
Dietary Fiber 1g	4%	4%
Sugars 10g		
Other Carbohydrate 14g		
Protein 1g		

CEREAL #2
Ingredients: Corn meal, sugar, whole wheat, modified corn starch, brown sugar syrup, partially hydrogenated soybean soil, honey, salt, nonfat milk, calcium carbonate, baking soda, dextrose, baking soda, dextrose, BHT.

Nutrition Facts
Serving Size 1 cup (30g)
Servings Per Container About 15

Amount Per Serving		with ½ cup skim milk
Calories	110	150
Calories from Fat	10	10
	% Daily Value**	
Total Fat 1g*	2%	2%
Saturated Fat 0g	0%	0%
Polyunsaturated Fat 0g		
Monounsaturated Fat 0g		
Cholesterol 0mg	0%	1%
Sodium 200mg	8%	11%
Potassium 85mg	2%	8%
Total Carbohydrate 24g	8%	10%
Dietary Fiber 3g	11%	11%
Sugars 6g		
Other Carbohydrate 15g		
Protein 3g		

CEREAL #3
Ingredients: Whole grain corn, oats, whole grain barley, whole grain rice, whole wheat, sugar, corn starch, corn bran, salt, calcium carbonate, partially hydrogenated soybean oil, trisodium phosphate, monoglycerides, vitamins and minerals.

1. Which cereal has the most dietary fiber? 3
2. Which cereal has the least dietary fiber? 2
3. Which cereal has the most sugar? 1
4. Which cereal has the least sugar? 3
5. How many teaspoons of sugar are in each cereal? #1 2 3/4 #2 2 1/2 #3 1 1/2
6. Which cereals contain whole grains? 1, 2, 3
7. Which cereal contains the least amount of whole grains? 2
8. The labels are from Golden Grahams, Honey Nut Cheerios, and Multi-Grain Cheerios. Identify each cereal by its number.
   Golden Grahams 2 Honey Nut Cheerios 1 Multi-Grain Cheerios 3

STUDENT WORKSHEET 3-2

TOPIC: Comparison of Artificial Sweeteners

Prepare 4 quarts of iced tea. (To prepare 1 quart, pour 4 cups of boiling water over 4 tea bags. Brew 4 minutes. Remove tea bags and chill tea in refrigerator for 2 hours or until cold. Repeat.) In this exercise, you will add 2 teaspoons of sugar, or its equivalent, to each cup of iced tea.

1. To the first quart of iced tea, add 8 teaspoons of sugar and mix thoroughly
2. To the second quart of iced tea, add 4 packets of Splenda (sucralose) and mix thoroughly
3. To the third quart of iced tea, add 4 packets of Equal (aspartame) and mix thoroughly
4. To the final quart of iced tea, add 4 packets of Sweet-One (acesulfame-K) and mix thoroughly

Next, prepare a blind taste testing of the sweetened iced teas by assigning each quart a number (see below) and labeling the number on the iced tea quart. Before pouring samples, make sure each product is thoroughly stirred. Ask students to fill out the evaluation form below.

Evaluation of Sweetened Iced Teas

1. As you taste each iced tea sample, please check off under the most appropriate category.

	Dislike Very Much	Dislike Slightly	Neither Like Nor Dislike	Like Slightly	Like Very Much
Sample #52					
Sample #35					
Sample #67					
Sample #12					

2. Do any of the samples leave any type of aftertaste? Which samples? What type of aftertaste (sweet and pleasant, metallic, artificial, etc.)

____

____

3. Rank each sample from your favorite to your least favorite.

#1	
#2	
#3	
#4	

4. Which sample do you think contains table sugar? ______________

STUDENT WORKSHEET 3-3

TOPIC: Bread Comparison: Whole Grain or Refined Grain?

Identify which bread is completely refined grain, which bread is completely whole grain, and which bread contains both refined and whole grains.

BREAD #1
INGREDIENTS: refined
Enriched Wheat Flour (Flour, Malted Barley Flour, Reduced Iron, Niacin, Thiamin Mononitrate (Vitamin B1), Riboflavin (Vitamin B2), Folic Acid), Water, High Fructose Corn Syrup, Yeast, Salt, Soybean Oil, Calcium Propionate (Preservative), Monoglycerides, Monocalcium Phosphate, Sodium Stearoyl, Lactylate, Calcium Sulfate, Soy Lecithin

BREAD #2
INGREDIENTS: both
Whole wheat flour, cornmeal, rolled oats, white bread flour, honey, oil, water, salt, yeast.

BREAD #3
INGREDIENTS: whole
Whole wheat flour, water, wheat gluten, high fructose corn syrup, contains 2% of less of: soybean oil, salt, molasses, yeast, mono and diglycerides, exthoxylated mono and diglycerides, dough conditioners (sodium stearoyl lactylate, calcium iodate, calcium dioxide), datem, calcium sulfate, vinegar, yeast nutrient (ammonium sulfate), extracts of malted barley and corn, dicalcium phosphate, diammonium phosphate, calcium propionate (to retain freshness).

CHAPTER 4 LIPIDS: FATS AND OILS

LEARNING OBJECTIVES

Upon completion of the chapter, the student should be able to:

1. Describe lipids and list their functions in foods and in the body
2. Describe the relationship between triglycerides and fatty acids
3. Define saturated, monounsaturated, and polyunsaturated fats and list foods in which each one is found
4. Describe trans fatty acids and give examples of foods in which they are found
5. Identify the two essential fatty acids, list their functions in the body, and give examples of foods in which they are found
6. Define cholesterol and lecithin, list their functions in the body, identify where they are found in the body, and give examples of foods in which they are found
7. Define rancidity
8. Describe how fats are digested, absorbed, and metabolized
9. Discuss the relationship between lipids and health conditions such as heart disease and cancer
10. State recommendations for dietary intake of fat, saturated fat, trans fat, monounsaturated fat, polyunsaturated fat, and cholesterol
11. Distinguish between the percentage of fat by weight and the percentage of kcalories from fat
12. Calculate the percentage of kcalories from fat for a food item
13. Discuss the nutrition and uses of milk, dairy products, and eggs on the menu

CHAPTER OUTLINE

1. **What are lipids?**

   Lipid is the chemical name for a group of compounds that includes fats, oils, cholesterol, and lecithin.

   Fats are usually solid at room temperature, oils are liquid. Fats are usually of animal origin, oils are usually of plant origin.

   Most of the lipids in foods, and also in the human body, are in the form of triglycerides.

2. **Functions of lipids**

   Fat accounts for 13 to 30% or more of a person's weight. Fat cells (adipose cells) can store loads of fat. The number of fat cells increases most during late childhood and early adolescence.

   Half of your fat cells are located just under the skin where fat provides insulation for the body, a cushion around critical organs, and optimum body temperature in the cold.

   Fat stores are a very compact way to store lots of energy, and fat spares protein from being burned for energy.

Fat also is present in all cell membranes and transports the fat-soluble vitamins through the body.

The essential fatty acids are needed for normal growth and development in infants and children. They are used to maintain the structural parts of cell membranes, and they play a role in the proper functioning of the immune system. From the essential fatty acids, the body makes hormone-like substances that control a number of body functions, such as blood pressure and blood clotting.

In foods, fats enhance taste, flavor, aroma, crispness, juiciness, tenderness, and texture. Fats also have satiety value.

3. **Triglycerides**
A triglyceride is composed of three fatty acids attached to glycerol.

Fatty acids differ from each other in terms of the length of the carbon chain and the degree of saturation. Short chain fatty acids have 6 or fewer carbons, medium chain fatty acids have 8 to 12 carbons, and long chain fatty acids have 14 to 20 carbons.

A fatty acid that contains only one point of unsaturation in the chain of carbons is called monounsaturated. A fatty acid that contains 2 or more points of unsaturation is called polyunsaturated. A fatty acid that is filled to capacity with hydrogens is called saturated.

Three types of triglycerides
1. A saturated triglyercide, also called a **saturated fat**, is a triglyceride in which most of the fatty acids are saturated.
2. A monounsaturated triglyceride, also called a **monounsaturated fat**, is a triglyceride in which most of the fatty acids are monounsaturated.
3. A polyunsaturated triglyercide, also called a **polyunsaturated fat**, is a triglyceride in which most of the fatty acids are polyunsaturated.

4. **Nutrition Science Focus: Lecithin and Triglycerides**
Lecithin is considered a phospholipid, a class of lipids that are like triglycerides except that one fatty acid is replaced by a phosphate group and choline or another nitrogen-containing group. Phospholipids are unique in that they are soluble in fat and water. Phospholipids such as lecithin are used by the food industry as emulsifiers in salad dressings, for example. Emulsifiers break up fat globules into small droplets, resulting in a uniform mixture than won't separate. Egg yolk are rich in lecithin and are used as emulsifiers in baking recipes. Lecithin is made in the liver so it is not essential.

Saturated fatty acids are filled to capacity with hydrogens (Fig 4-2). When a hydrogen is missing from two neighboring carbons, a double bond forms between the carbon atoms, and the fatty acid is called unsaturated. A fatty acid that contains only one point of unsaturation it is called monounsaturated. If the chain has 2 or more points of unsaturation, the fatty acid is called polyunsaturated.

5. **Triglycerides in foods**
   All food fats contain a mixture of saturated and unsaturated fats.

   1. Fruits and vegetables – most have no fat; exceptions are avocado, olives and coconuts
   2. Breads, cereals, rice, pasta, and grains – most are low in fat except for croissants, biscuits, cornbread, and some granolas and crackers; most baked goods such as cakes and pies are quite high in fat
   3. Dry beans and peas, nuts, and seeds – most beans and peas are low in fat, nuts and seeds are quite high in fat – but most of the fat is monounsaturated and polyunsaturated
   4. Meat, poultry, and fish – meat tends to have more fat than poultry which tends to have more fat than fish; within each group there are high fat and lower in fat choices
   5. Dairy – most regular dairy products are high in fat; dairy products such as no fat or reduced fat are healthier choices
   6. Fats, oils, and condiments – fats and oils are almost all fat

   - Saturated fat is found in cheese, beef, whole milk and other full-fat dairy products, many fats in baked goods, margarine, butter, and tropical oils. Animal fat tends to contain at least 50% saturated fat.
   - Monounsaturated fats include olive oil, canola oil, and peanut oil.
   - Polyunsaturated fats are found in greatest amounts in safflower, corn, soybean, sesame, and sunflower oils. These oils are commonly used in salad dressings and as cooking oils. Nuts and seeds also contain polyunsaturated fats, enough to make nuts and seeds a rather high-kcalorie snack food depending on serving size.

6. **Trans fats**
   Trans fatty acids are made during hydrogenation of liquid fats to solid fats, such as to make vegetable shortening and margarine. During hydrogenation, some unsaturated fatty acids lose their natural bend and become straight like saturated fatty acids. This makes them more solid, but because they are straight, they also behave like saturated fats in the body (in other words, they increase the blood levels of LDL cholesterol). Hydrogenated fats can be found in some baked goods, snack foods, French fries, margarines/shortenings, and chips/popcorn.

   On food labels you will find the amount of trans fat per serving on its own line under "Total Fat."

7. **Essential fatty acids: linoleic acid and alpha-linolenic acid**
   Linoleic acid is an omega-6 fatty acid found in vegetable oils (corn, safflower, soybean, cottonseed, and sunflower), margarines, and salad dressings.

   Alpha-linolenic acid is the leading omega-3 fatty acid found in food, and it is found in canola, flaxseed, soybean, and walnut oils. Other good sources include ground flaxseed, walnuts, and soy products.

   The body converts alpha-linolenic into DHA and EPA (also omega-3s). DHA and EPA are found in fatty fish such as salmon and mackerel. DHA and EPA are excellent for heart health; they reduce blood pressure, blood clots, heart rate, and blood triglyceride levels.

Both EFAs services as part of cell membranes, play a role in the proper functioning of the immune system, and are vital to normal growth and development in infants and child. DHA and EPA are especially important for proper brain and eye development during pregnancy and infancy.

Americans get more than enough linoleic, but not enough alpha-linolenic acid. The ratio of omega-3 to omega-6 fatty acids in the diet is important in regulating your blood pressure, blood clotting, and inflammation. Having a healthy ratio of omega-3 to omega-6 fatty acids seems to lower blood pressure, prevent blood clot formation, and reduce inflammation.

8. **Cholesterol**
Cholesterol is the most abundant sterol, a category of lipids.

Cholesterol is present in every cell in your body. It is needed to make bile acids, cell membranes, many hormones (such as sex hormones), and vitamin D. Cholesterol builds up in the plaque that clogs arteries in your body. High blood cholesterol is a risk factor for heart disease.

Cholesterol is found only in foods of animal origin: egg yolks (not egg whites), meat, organ meats, poultry, fish, milk, and milk products. Lower-fat milk products contain less cholesterol than full-fat milk products.

We take in about 200 to 400 milligrams of cholesterol daily and the liver and body cells also make cholesterol (about 700 milligrams), therefore it is not an essential nutrient.

9. **Digestion, absorption, and metabolism of fats**
Because fat and water do not mix, fat digestion is difficult without emulsifiers. Lingual lipase, an enzyme made in the salivary glands, has a minor role in fat digestion in adults and an important role in infants. In the stomach gastric lipase breaks down mostly short-chain fatty acids.

With the help of bile and various enzymes, fats are broken down into monoglycerides, fatty acids, and glycerol, so they can be absorbed across the intestinal wall. Once absorbed into the cells of the small intestine, triglycerides are reformed.

Because fats would float in clumps in either the blood or lymph, the body makes lipoproteins. Chylomicrons, a type of lipoprotein, carry triglycerides and cholesterol from the intestines through the lymph system to the bloodstream. In the bloodstream an enzyme, lipoprotein lipase, breaks down the triglycerides into fatty acids and glycerol so they can be absorbed into the body's cells.

The primary sites of lipid metabolism are the liver and the fat cells. The liver manufactures triglycerides and cholesterol. These products are carried through the body by the liver's equivalent of chylomicrons: very low density lipoprotein (VLDL). VLDLs release triglycerides, with the help of lipoprotein lipase, throughout the body. VLDLs are converted in the blood into LDLs.

Low density lipoproteins, often called "bad cholesterol," are mostly made of cholesterol and transport much of the blood cholesterol to the body's cells. The LDLs not absorbed by cells are somehow involved in depositing cholesterol on the inner blood vessel wall, causing hardening and narrowing of the arteries.

High density lipoprotein, also called "good cholesterol," travels throughout the body picking up cholesterol. HDL returns the cholesterol to the liver to be disposed.

10. **Nutrition Science Focus: Lipoproteins**
    Lipoprotein lipase is an enzyme in the bloodstream that breaks down the triglycerides in the chylomicrons in fatty acids and glycerol so they can be absorbed into the body's cells. The cells can use the fatty acids for energy, which the muscle cells often do, or make triglycerides for storage, which fat cells often do.

    The primary sites of lipid metabolism are the liver and the fat cells. The liver makes triglycerides and cholesterol that are carried through the body by VLDL. Once most triglycerides are removed from VLDL with the help of lipoprotein lipase, VLDL is converted in the blood into LDL. LDL distributes cholesterol, triglycerides, and phospholipids to the body's cells. Certain cells (especially in the liver) can absorb the entire LD particle—they play an important role in the control of blood cholesterol concentrations.

11. **Lipids and health**
    Heart disease is the #1 killer of both men and women in the U.S. Too much cholesterol leads to accumulation of cholesterol-laden plaque in blood vessel linings—called atherosclerosis. Aterosclerosis can lead to:
    - Coronary heart disease and myocardial infarction—when a blood clot completely obstructs a coronary artery (heart attack)
    - Stroke—Damage to brain cells resulting from an interruption of blood flow to the brain.

    The primary ways in which LDL cholesterol levels become too high is through eating too much saturated fat, trans fat, and to a lesser extent cholesterol. Dietary factors that lower LDL cholesterol include replacing saturated and trans fats with polyunsaturated and monounsaturated fats, and to a lesser extent soluble fiber and soy protein.

    Being overweight and inactive promotes high blood cholesterol levels. Age, gender, and heredity also can affect your blood cholesterol levels, but you can't control these factors as you can the other factors mentioned. You can reduce your risk by adopting some dietary changes, losing weight, exercising, or quitting smoking.

    Fat may be involved in certain cancers, such as prostate cancer.

12. **Dietary recommendations**
The DRI do not include an AI or RDA for total fat (except for infants) because there is not enough scientific data to determine an appropriate level. AMDR for fat was set:

1 to 3 years old	30–40% of kcalories
4 to 18 years old	25–35%
Over 18 years old	20–35%

No AI nor RDA was set for saturated fat or cholesterol because they are made in the body and have no known role in preventing chronic diseases. Likewise no AI or RDA was set for trans fatty acids because they are not essential and provide no known benefit to health.

The Food & Nutrition Board recommends keeping intake of saturated fat, cholesterol, and trans fat as low as possible.

In 2002 an AI was set for EFAs.

The Dietary Guidelines for Americans (2005) and the American Heart Association recommend a diet for healthy American that:

- provides less than 7–10% of total kcalories from saturated fat
- provides less than 1% of total kcalories from trans fat
- provides less than 300 milligrams per day of cholesterol
- replaces most saturated fats with sources of polyunsaturated and monounsaturated fatty acids, such as fish, nuts, and vegetable oils

(These recommendations do not apply to children under 2 years of age.)

Explain the difference between the percentage of fat by weight (for example, roast beef is 90% fat free) and the percentage of kcalories from fat (45% of the kcalories in one serving of roast beef are from fat). To find out the percentage of kcalories from fat in any serving of food, simply divide the number of kcalories from fat by the number of total kcalories and then multiply the answer by 100.

13. **Ingredient Focus: Milk, Dairy Products, and Eggs**
Milk is a good source of protein, carbohydrate, riboflavin, vitamins A and D. Describe types of fluid milk: whole, reduced fat, low-fat, fat-free, buttermilk, eggnog, and lactase-treated milk. When cooking with milk, use a moderate heat and heat to the milk slowly (but not too long) to avoid curdling and scorching.

Cheese is an excellent source of nutrients such as protein and calcium. Cheeses made from whole milk or cream are also high in saturated fat.

Eggs are high in protein and cholesterol (215 milligrams cholesterol/egg). They also contain varying amounts of many vitamins and minerals.

14. Culinary Science

Cream is the fat in milk that is used to make butter, cream for whipping and other products. To make whipped cream, you will need a cream with at least 30% fat. If you whip cream for too long, you will get butter and buttermilk. Cheese is made by curdling milk, stirring and heating the curd, draining off the whey, collecting and pressing the curd, and in some cases ripening. The curds are rich in protein and fat. The amount of fat in cheese depends on the fat in the milk it was made from.

When using cheese in sauces and soups, avoid using one that becomes stringy when cooked, such as mozzarella and cheddar. Add the cheese as late as possible in the cooking to prevent the protein from becoming hard and squeezing out the fat. Starch ingredients help to stabilize the cheese during cooking.

Like cheese eggs are full of protein and should not be cooked at high temperatures and long cooking times because the egg whites will shrink and become tough and rubbery.

Rancidity is the deterioration of fat, resulting in undesirable flavors and odors.

15. Food Facts: Oils and Margarines

Table 4-16 gives information on various oils.

Margarine must contain vegetable oil and water and/or milk or milk solids. It is made by hydrogenation. Margarine and butter must contain at least 80% fat by weight. Margarines vary according to: their physical form, type of vegetable oil(s) used, percent fat by weight, and nutrient profile. catogorized by how they are pressed

Besides oils, margarine, and butter, you can buy: blends, butter-flavored buds, and margarine containing ingredients that reportedly lower blood cholesterol levels.

16. **Hot Topic: Trans Fats in Restaurants**

Restaurants in certain areas, such as New York City, have been told to stop serving foods with significant amounts of trans fats. Restaurants have been doing so by changing their oils for cooking and frying, choosing healthy spreads, and ordering foods without trans fats.

Laws such as those in New York City are controversial, with different organizations taking varying positions. The National Restaurant Association supports gradually phasing out trans fat in restaurant foods; however, they oppose inflexible bans with unrealistic timetables. Some health related groups want more cities to take similar actions.

NUTRITION WEB EXPLORER

The American Heart Association: www.deliciousdecisions.org
Visit the nutrition site for the American Heart Association and look at the recipes in their cookbooks. Write down three cooking methods, three seasonings, and three cooking substitutions that are heart-healthy.

Cabot Cheese: www.cabotcheese.com
Cabot makes some excellent cheeses that are lower in fat. Click on "Our Products" and then click on "Reduced-Fat Cheddar." List the products you find. Hopefully you might be able to taste them in class!

Trans Fat Help Center: www.notransfatnyc.org
Visit this website that was designed to help New York City food service establishments. Write two paragraphs on their advice for baking and frying without trans fat.

Golden Valley Flax: www.flaxhealth.com/recipes.htm
Flaxseed is a wonderful source of omega-3 fatty acids. The body cannot break down whole seeds to access the omega-3 containing oil, so most recipes use ground seeds. Read the recipes on this web page and try one out. What is the flavor of flaxseed like? Can you put it on cereal or a salad?

CHAPTER REVIEW QUIZ

Key Terms: Matching

1. Lipids
2. Fat
3. Oil
4. Triglyceride
5. Essential fatty acids
6. Fatty acids
7. Glycerol
8. Saturated fatty acid
9. Unsaturated fatty acid
10. Point of saturation
11. Monounsaturated fatty acid
12. Polyunsaturated fatty acid
13. Trans fats
14. Hydrogenation
15. Rancidity
16. Cholesterol
17. Lecithin
18. Monoglycerides

a. The deterioration of fat, resulting in undesirable flavors and odors
b. Triglycerides with only one fatty acid
c. Fatty acids that the body cannot produce, making them necessary in the diet
d. A fatty acid that contains two or more double bonds in the chain
e. A lipid that is usually liquid at room temperature
f. Unsaturated fatty acids that lose a natural bend or kink so that they become straight after being hydrogenated
g. A derivative of carbohydrate that is part of triglycerides
h. The major form of lipid in food and in the body
i. A process in which liquid vegetable oils are converted into solid fats by the use of heat, hydrogen, and certain metal catalysts
j. A group of fatty substances that are soluble in fat, not water, and that provide a rich source of energy and structure to cells
k. A fatty acid that is filled to capacity with hydrogens
l. Major component of most lipids. Three fatty acids are present in each triglyceride
m. A phospholipids and a vital component of cell membranes that acts as an emulsifier
n. A fatty acid with at least one double bond
o. The most abundant sterol that is present in every cell in your body
p. The location of the double bond in unsaturated fatty acids
q. A lipid that is solid at room temperature
r. A fatty acid that contains only one double bond in the chain

Multiple Choice

1. Glycerol is a derivative of:
   a. Carbohydrate
   b. Lecithin
   c. Protein
   d. None of the above

2. Lipids include:
   a. Carbon
   b. Phosphorus
   c. Hydrogen
   d. a and c

3. The food with the largest amount of saturated fat is:
   a. 1 Tbsp. canola oil
   b. 1 Tbsp. stick margarine
   c. 1 Tbsp. margarine spread
   d. 1 Tbsp. vegetable shortening

4. Great sources of monounsaturated fats include:
   a. Ground beef
   b. Safflower oil
   c. Sesame oil
   d. Peanut oil

5. Americans generally get plenty of this lipid in their diet:
   a. Linoleic acid - omega 6
   b. Linolenic acid omega 3
   c. DHA
   d. EPA

6. Cholesterol is not found in the following food:
   a. Non-fat milk
   b. Egg yolk
   c. Egg white
   d. Tuna

7. The primary site for lipid metabolism is:
   a. Small intestine
   b. Stomach
   c. Fat cells
   d. Large intestine

8. The Dietary Guidelines for Americans (2005) recommend a diet that provides:
    a. Less than 100 mg per day of cholesterol
    b. Less than 200 mg per day of cholesterol
    c. Less than 300 mg per day of cholesterol
    d. Less than 400 mg per day of cholesterol

9. The condition in which an accumulation of cholesterol-laden plaque lines the blood vessels is known as:
    a. Angina
    b. Myocardial infarction
    c. Stroke
    d. Atherosclerosis

10. A substance that emulsifies fat is:
    a. Gastric lipase
    b. Chylomicron
    c. Bile
    d. Lingual lipase

True/False

1. Essential fatty acids are necessary for proper functioning of the respiratory system.
    a. True b. False

2. All food fats, animal or vegetable, contain a mixture of saturated and unsaturated fats.
    a. True b. False

3. Trans fatty acids occur naturally at low levels in meat and dairy foods.
    a. True b. False

4. The body converts linoleic into docosahexaenoic acid and eicosapentaenoic acid.
    a. True b. False

5. Cholesterol is present in every cell in the body.
    a. True b. False

6. Lecithin is an essential nutrient.
    a. True b. False

7. Fat can cause several types of cancer.
    a. True b. False

8. There is no RDA or AI for fat (except for infants), saturated fat, cholesterol, or trans fatty acids.
    a. True b. False

9. Heredity, unlike age and gender, does not determine the amount of cholesterol that the body makes.
   a. True b. False

10. Heart disease is the leading cause of death in the United States.
   a. True b. False

Short Answer

1. What are the three types of triglycerides?

   Saturated fatty acids, monounsaturated fatty acids, and polyunsaturated fatty acids

2. The spot where a double bond is located in a fatty acid is referred to as a point of unsaturation.

3. Which phospholipid is a vital component of cell membranes?

   lecithin

4. Why is HDL considered good cholesterol?

   It travels through the body picking up the cholesterol, which it brings back to the liver for breakdown & disposal - it prevents cholesterol buildup in the arterial walls

5. What is the function of a chylomicron?

   Responsible for carrying mostly triglycerides & some cholesterol, through the intestines through the lymph system to the blood stream

STUDENT WORKSHEET 4-1

TOPIC: Recipe Modification

Suggest how each of these recipes can be modified to contain less fat and/or less saturated fat.

Recipe	Directions
1. ***Crab Cakes*** 5 pound crabmeat, flaked ¼ cup Dijon mustard ¼ cup horseradish ½ cup chives 1 1/2 cups Japanese bread crumbs 1 1/4 cups mayonnaise Salt and pepper as desired Vegetable oil	1. Combine the crabmeat and other ingredients in a bowl. 2. Mix well. 3. Fry the crabcakes in oil in a skillet until browned and cooked through, about 3 minutes on each side.
2. ***Au Gratin Potatoes*** 2 pounds potatoes, peeled and sliced 1 large onion, minced 2 tablespoons butter 1 pint grated Gruyere cheese 2/3 cup heavy cream Salt and pepper to taste	1. Steam or boil potatoes until just tender. 2. Place the potatoes in a greased pan. 3. Sauté the onion in butter. 4. Add the cheese and stir until cheese is melted. 5. Add the heavy cream, salt, and pepper. 6. Pour the sauce over the potatoes and bake at 350°F oven for 20–30 minutes until heated through.
3. ***Blueberry Cobbler*** 1 cup all-purpose flour 2 tablespoons sugar 1½ teaspoons baking powder ½ teaspoon ground cinnamon ¼ cup butter ½–2/3 cup sugar 1 tablespoon cornstarch ¼ cup water 4 cups fresh or frozen blueberries 1 egg ¼ cup milk vanilla ice cream	1. For topping, in a medium bowl stir together flour, 2 tbsp. sugar, baking powder, and cinnamon. Cut in butter until mixture resembles crumbs. 2. For filling, in a saucepan combine 1/2 to 2/3 cup sugar, 1 tablespoon cornstarch, and ¼ cup water. Stir in 4 cups blueberries. 3. Cook and stir until thickened and bubbly. Place in a 2-quart square baking dish. 4. Stir the egg and milk in a small bowl, then add to flour mixture, stirring just to moisten. 5. Using a spoon, drop the topping into small mounds on the blueberry filling. 6. Bake at 400°F for 20 to 25 minutes or until a toothpick is inserted and comes out clean. Serve with ice cream.

STUDENT WORKSHEET 4-2

TOPIC: Name That Fat Substitute!

Following are ingredient listings from four products made with fat substitutes. Using "Hot Topics: Fat Substitutes" on page 177 as a guide, identify the fat substitutes in these foods.

Creme-Filled Chocolate Cupcakes—0 grams fat/cupcake
Sugar, water, corn syrup, bleached flour, egg whites, nonfat milk, defatted cocoa, invert sugar, modified food starch (corn, tapioca), glycerine, fructose, calcium carbonate, natural and artificial flavors, leavening, salt, dextrose, calcium sulfate, oat fiber, soy fiber, preservatives, agar, sorbitan monostearate, mono- and diglycerides, carob bean gum, polysorbate 60, sodium stearoyl lactylate, xanthan gum, sodium phosphate, maltodextrin, guar gum, pectin, cream of tartar, sodium aluminum sulfate, artificial color.

Low-Fat Mayonnaise Dressing—1 gram fat per tablespoon
Water, corn syrup, liquid soybean oil, modified food starch, egg whites, vinegar, maltodextrin, salt, natural flavors, gums (cellulose gel and gum, xanthan), artificial colors, sodium benzoate and calcium disodium EDTA.

Lite Italian Dressing—0.5 grams fat/2 tablespoons
Water, distilled vinegar, salt, sugar, contains less than 2% of garlic, onion, red bell pepper, spice, natural flavors, soybean oil, xanthan gum, sodium benzoate, potassium sorbate and calcium disodium EDTA, yellow 5 and red 40.

Light Cream Cheese—5 grams fat per 2 tablespoons
Pasteurized skim milk, milk, cream, contains less than 2% of cheese culture, sodium citrate, lactic acid, salt, stabilizers (xanthan and/or carob bean and/or guar gums), sorbic acid, natural flavor, vitamin A palmitate.

STUDENT WORKSHEET 4-3

TOPIC: Comparison of Fat in Dairy Products

1. Using food labels, Appendix A in your textbook, or an Internet website that offers nutrient analysis (such as http://www.ars.usda.gov/main/site_main.htm?modecode=12354500), fill in the following table.

Dairy Product	Kcalories/cup	Fat/cup	Sat Fat/cup	Cholesterol/cup
Whole Milk				
Reduced Fat Milk (2% fat)				
Low Fat Milk (1% fat)				
Fat-Free milk (skim)				
Ice Cream, vanilla				
Reduced fat ice cream, vanilla				
Frozen Yogurt, vanilla				

2. Answer the following questions.

Which milk has the most fat? ____________________

Which milk has the least fat? ____________________

Does fat-free milk contain less saturated fat than reduced-fat or regular milk?____________________

Does fat-free milk contain less cholesterol than reduced-fat or regular milk? ____________________

How many kcalories do you save by drinking fat-free rather than regular milk?____________________

Which frozen dessert tends to be lowest in fat and saturated fat? ____________________

STUDENT WORKSHEET 4-4

TOPIC: Baking with Different Fats and Comparing Them Nutritionally

1. Make this recipe with shortening. Then make 2 modifications: substitute 1 cup of margarine for the shortening, and then substitute 2/3 cup soybean oil for the shortening.

Snickerdoodles

1 cup shortening	2 3/4 cups all-purpose flour
(or 1 cup margarine, or 2/3 cup soybean oil)	2 tsp. cream of tartar
1 1/2 cups granulated sugar	1 tsp. soda
2 eggs	1/4 tsp. salt
	2 tablespoons sugar plus 2 teaspoons cinnamon

Preheat oven to 400 degrees. Cream shortening or margarine (or mix oil) with sugar. If using oil, stir only until blended—do not overmix. Add eggs. Mix only until blended. Add remaining ingredients and stir only until well blended. Roll dough into balls the size of a small walnut and roll each ball in sugar/cinnamon mixture. Place on ungreased cookie sheets and bake until lightly browned for 8 to 10 minutes. Yields about 50 three-inch cookies.

2. Describe each of the cookies.

Cookie	Color	Texture	Taste
Shortening			
Margarine			
Soybean oil			

Rank the cookies in order of overall taste and quality. ______________________________

3. Compare the total fat, saturated fat, trans fat, and monounsaturated fat in 1 cup of shortening, 1 cup of 80% stick margarine, and 2/3 cup of soybean oil using Figure 4-17 on page 182 in your textbook.

	Total Fat	Saturated Fat	Trans Fat	Mono Fat	Cholesterol
1 cup shortening					
1 cup margarine-stick, 80% fat					
2/3 cup soybean oil					

Which cookie had the best nutritional profile? the second-best nutritional profile?

STUDENT WORKSHEET 4-5

TOPIC: Exploring Flavor and Cost Using Different Oils in Sautéing

Directions: Cut skinless, boneless chicken breasts into ½-inch dice. Sauté with 1 of the following oils using these directions.
1 fl oz canola oil or peanut oil or olive oil

1. Heat the oil in a large sauté pan or skillet until very hot. Add the chicken and sauté rapidly until no longer pink. If the chicken sticks to the pan, use a spatula to stir.

2. Put the chicken on a plate.

3. Continue to sauté more chicken with the other oils.

Now fill in the following chart with your comments on flavor as well as the cost of 1 oz of oil and smoking point.

	Taste	Cost/oz.	Smoking Point
Canola oil			
Olive oil			
Peanut oil			

Which product was your favorite?

CHAPTER 5 PROTEIN

LEARNING OBJECTIVES

Upon completion of the chapter, the student should be able to:

1. Identify and describe the building blocks of protein
2. List the functions of protein in the body
3. Explain how protein is digested, absorbed, and metabolized
4. Distinguish between complete protein and incomplete protein and list examples of foods that contain each
5. Explain the potential consequences of eating too much or too little protein
6. State the dietary recommendations for protein
7. Discuss the nutrition and uses of meat, poultry, and fish on the menu
8. Describe soy products, their health benefits, and how to use them on the menu
9. Discuss the advantages and disadvantages of irradiation

CHAPTER OUTLINE

1. **What is protein?**

   Proteins are an essential part of all living cells found in animals and plants. Whereas carbohydrates and lipids are used primarily for energy, proteins function in a very broad sense to build and maintain your body.

2. **Protein's structure**

   Like carbohydrate and fat, protein contains carbon, hydrogen, and oxygen; but unlike these other nutrients, protein contains the chemical element nitrogen. Proteins are a major source of nitrogen for the body.

   Amino acids are the building blocks of protein. Proteins are strands of amino acids. There are 20 different amino acids, each consisting of a backbone to which a side group is attached. The amino acid backbone is the same for all amino acids, but the side group varies.

   Nine of the amino acids are essential or indispensable. The remaining are called nonessential amino acids. Under certain circumstances, one or more nonessential amino acids may become essential, which is referred to as conditionally essential.

   When the amino-acid backbones join end-to-end to form a protein, the bonds are called peptide bonds. Proteins often contain 35 to several hundred or more amino acids. Protein fragments with 10 or more amino acids are called polypeptides.

3. **Functions of protein**

   Proteins function as part of the body's structure—skin, hair, bones, muscles, blood vessels, blood, part of DNA.

   Proteins are used to build and maintain body tissues.

   Many enzymes (catalysts) and hormones (chemical messengers) are made of protein.

Enzymes contain a special pocket called the active site. Various substances fit into the pocket, undergo a chemical reaction, and then exit the enzymes in a new form. Enzymes do this without being changed themselves. Enzymes help break down substances such as food you eat, build up substances such as bone, and change one substance into another (such as glucose into glycogen).

Hormones regulate certain body activities so that a constant internal environment (called homeostasis) is maintained.

All antibodies (proteins in the blood that bind with foreign bodies) are protein. Antibodies are vital to your immune response, in other words, how your body responds to foreign substances such as a virus.

Protein acts as taxicabs in the body, transporting iron, fats, minerals, and oxygen.

Protein plays a role in maintaining fluid and acid-base balance. Foods that you eat as well as the normal processes that go in your body produce acids and bases. It is crucial that you blood not build up high levels of acids or bases. Some proteins in your blood have the ability to neutralize both acids and bases.

Protein provides energy as last resort.

Protein is also important for blood clotting. Protein fibers known as fibrin help form a clot so bleeding stops.

4. **Nutrition Science Focus: Proteins**
   Each protein has its own characteristics primary structure (number and sequence of amino acids), secondary structure (such as coiling), and tertiary structure (folding and twisting), which make it functional.

   The instructions to make proteins in your body reside in the core, or nucleus, of each of your body's cells. In the nucleus are molecules called DNA which contain vital genetic information. Every human cell (except for mature red blood cells) contains the same DNA. DNA exist as 2 long, paired strands that are spiraled into a double helix. Each strand is made up of millions of chemical building blocks called bases. There are only 4 different bases in DNA, but they can be arranged and rearranged in countless ways. The order in which the bases occur determines the messages to be conveyed, such as how proteins are made. Each DNA molecule in the nucleus is housed in a chromosome. You have 23 sets of chromosomes in each of your cells. Genes, segments of DNA, carry particular set of instructions that allows a cell to produce a specific product such as an enzyme.

5. **Digestion, absorption, and metabolism**
   Protein digestion takes place in the stomach and small intestine. The hydrochloric acid in the stomach converts pepsinogen to the enzyme pepsin which splits peptide bonds in proteins.

   Next, in the small intestine, proteases split up proteins into short peptide chains and amino acids. The brush border of the small intestine produces several peptidases, enzymes that break down short peptide chains into amino acids, dipeptides, and tripeptides.

When tripeptides and dipeptides enter the intestinal cells, they are split into amino acids. Because amino acids are water soluble, they travel easily in the blood to the liver and then to the cells that need them.

An amino acid pool in the body provides the cells with a supply of amino acids for making protein. If the body is making a protein and can't find an essential amino acid for it, the protein can't be competed and the partially completed protein is taken apart.

6. **Protein in food**
   Protein is found in animal and plant foods. Protein is highest in animal foods.

   Of plant foods, grains, legumes, and nuts usually have more protein than vegetables and fruits.

   Animal foods generally are higher in fat and saturated fat than plant foods (which have no cholesterol).

   Complete protein provide all of the essential amino acids in the proportions needed—examples include animal protein sources such as milk.

   Incomplete protein contains at least one limiting amino acid—examples include most plant proteins. Some plant proteins, such as quinoa and protein made from soysbeans (isolated soy protein), are complete proteins.

   Complementary proteins, such as peanut butter on whole wheat bread, each supply the limiting amino acid lacking in the other food.

   Plant proteins, although incomplete, are not low in quality. In the right amounts and combinations, plant proteins support growth and maintenance.

7. **Protein and health**
   There is no benefit to eating too much protein. Eating too much high-fat animal protein can increase your blood cholesterol, calcium loss from the body (if calcium intake is low), and worsen kidney problems in people with renal disease. High intakes of animal proteins are also associated with certain cancers, such as cancer of the colon. The American Cancer Society advises you to limit consumption of red meats, especially ones high in fat and cholesterol.

   Eating too little protein can cause problems—such as slowing down the protein rebuilding and repairing process and weakening the immune system.

   Protein-energy malnutrition (PEM) refers to a broad spectrum of malnutrition form mild to serious. Kwashiorkor, a type of PEM, is usually seen in children who are getting inadequate amounts of protein and only marginal amounts of kcalories. Kwashiorkor is characterized by retarded growth and development and a protruding abdomen due to edema.

Marasmus, another type of PEM, is characterized by severe insufficiency of kcalories and protein, which accounts for the child's gross underweight, lack of fat stores, and wasting away of muscles.

8. **Dietary recommendations**
RDA for men and women is 0.8 g/kg or 0.36 grams per pound of body weight. The amount of protein needed daily is proportionally higher during periods of growth because the body is in a state of positive nitrogen balance.

Positive nitrogen balance is a condition in which the body excretes less protein than is taken in—this can occur during growth and pregnancy. Negative nitrogen balance is a condition in which the body excretes more protein than is taken in—this can occur during starvation and certain illnesses.

Most Americans eat more protein than the RDA.

AMDR for protein for adults is 10 to 35% of total kcalories.

9. **Meats, poultry, and fish**
Nutrition: In comparison to red meats, skinless white meat chicken and turkey are lower in total fat and saturated fat, and comparable in cholesterol. Most fish is lower in fat, saturated fat, and cholesterol than are meat and poultry.

Meat is a good source of many important nutrients, including protein, iron, copper, zinc, and some of the B vitamins such as B6 and B12.

Fish and shellfish are excellent sources of protein and are relatively low in kcalories.

Steps to prepare healthy meat, poultry, and seafood:
1. Select a lean cut.
2. Use flavorful rubs and marinades.
3. Choose a cooking method that will provide a flavorful, moist product and that adds little or no fat to the food.
4. Think of how to flavor the dish (i.e. herbs and spices, smoking, etc.)
5. Fish must be cooked very carefully and not overdone. Serve immediately.

10. **Food Facts: Soy Foods and Their Health Benefits**
Various forms of soybeans are eaten in the US: soy oil, tofu(bean curd), soy sauce, miso, tempeh, textured soy protein. meat analogs, soy cheese, soy flour, edamame, natto, soymilk, and soy nuts.

Much research is being done on the health effects of soy. Foods containing soy protein may reduce the risk of coronary heart disease when consumed as part of a diet low in saturated fat and cholesterol. Soybeans contain phytoestrogens, which are chemically similar to estrogen, the female hormone. Some possible health effects of soy are due to the fact that phytoestrogens can mildly mimic the actions of estrogen in the body. Some studies suggest that soyfoods may reduce hot flashes in women after menopause when natural estrogen is lacking, as well as prevent bone loss. Soy's possible role in preventing breast cancer is uncertain.

11. Irradiation

In 1997 the FDA approved the treatment of red-meat products with a measured dose of radiation in order to control *E. coli* 0157. The same irradiation process is used to sterilize medical products such as bandages. Many spices in the US are irradiated.

Irradiation is a cold process that gives off little heat. It interferes with bacterial genetics so the organisms can no longer survive or multiply.

Some restaurant chains are using irradiated beef and the U.S. government is irradiating ground beef that is given to the School Lunch program. Irradiated foods must be labeled.

NUTRITION WEB EXPLORER

Nutrition Data: www.nutritiondata.com
On the home page for Nutrition Data, you can click on the name of many fast-food chains to get a nutrient analysis of their foods. Compare and contrast the nutritional values (kcalories, fat, saturated fat, protein, cholesterol) for five entrees for at least one restaurant chain. Pick entrées that range from being very high in fat to being much lower.

Cattleman's Beef Board: www.beefitswhatsfordinner.com
Click on "Land of Lean Beef" and write down the names of at least ten lean cuts of beef.

Bass on Hook: www.bassonhook.com/fishforfood/fishcookingtechniques.html
Read "Fish Cooking Techniques." How long does it take to cook fresh fish that is 1 inch thick? Frozen fish that is 1 inch thick? What is the difference between blackening and bronzing?

Georgia Eggs: www.georgiaeggs.org

Click on "Eggs A to Z" and then click on "Egg Products." List three different types of egg products. What are the advantages of using these products? Click on "Nutrition" and find out why eggs are a high quality protein.

CHAPTER REVIEW QUIZ

Key Terms: Matching

1. Proteins
2. Amino acids
3. Essential amino acids
4. Peptide bonds
5. Enzymes
6. Hormones
7. Antibodies
8. Antigens
9. Acid-base balance
10. Acidosis
11. Alkalosis
12. Fibrin
13. Denaturation
14. Pepsin
15. Proteases
16. Kwashiorkor
17. Marasmus

a. A dangerous condition in which the blood is too basic
b. A type of PEM associated with children who are getting inadequate amounts of protein and only marginal amounts of kcalories
c. Chemical messengers in the body
d. Enzymes that break down protein
e. Major structural parts of the body's cells that are made of nitrogen-containing amino acids assembled in chains
f. Proteins in the blood that bind with foreign bodies or invaders
g. Protein fibers involved in forming clots so that a cut or would will stop bleeding
h. The building blocks of protein
i. A type of PEM characterized by severe insufficiency of kcalories and protein that accounts for the child's gross underweight and wasting away of muscles
j. Foreign invaders in the body
k. Amino acids that either cannot be made in the body or cannot be made in the quantities needed by the body
l. The principle digestive enzyme of the stomach
m. The bonds that form between adjoining amino acids
n. The process by which the body buffers the acids and bases normally produced in the body so that the blood is neither too acidic nor too basic
o. A process in which a protein uncoils and loses its ability to function
p. Catalysts in the body that help break down substances, build up substances, and change one substance into another
q. A dangerous condition in which the blood is too acidic

True/False

1. The amino acid backbone is the same for all amino acids, but the side group varies to make each amino acid unique.
   a. True   b. False

2. Proteins often contain 10 to 35 amino acids.
   a. True   b. False

3. Skin lives for 3 to 7 days.
   a. True   b. False

4. Hormones regulate body activity so that homeostasis is maintained.
   a. True   b. False

5. When tripeptides and dipeptides enter the stomach cells, they are split into amino acids in order to travel more easily in the blood to the liver.
   a. True   b. False

6. Food proteins that provide all the essential amino acids in proportions needed by the body are complete proteins.
   a. True   b. False

7. Only plant proteins are complete proteins.
   a. True   b. False

8. Eating too much protein will have no health benefits such as bigger muscles or increased immunity.
   a. True   b. False

9. High protein intake is associated with certain cancers such as colon cancer.
   a. True   b. False

10. Negative nitrogen balance usually occurs during pregnancy.
    a. True   b. False

Multiple Choice

1. All of the following amino acids are essential except:
   a. Histidine
   b. Arginine
   c. Lysine
   d. Leucine

2. Which structure of protein is responsible for folding and twisting?
    a. Primary structure
    b. Secondary structure
    c. Tertiary structure
    d. Quaternary structure

3. Which functions are enzymes responsible for in the body?
    a. Build up substances
    b. Change one substance into another
    c. Break down substances
    d. All of the above

4. Proteins function as part of the body's structure in:
    a. Hair
    b. Bones
    c. The digestive tract
    d. All of the above

5. Protein digestion takes place in the:
    a. Stomach
    b. Liver
    c. Large Intestine
    d. Fat cells

6. What is the 2002 RDA for protein for healthy adults?
    a. 0.36 gram per pound of body weight
    b. 0.45 gram per pound of body weight
    c. 0.80 gram per pound of body weight
    d. 1.0 gram per pound of body weight

7. Red meat and skinless white meat have similar amounts of:
    a. B vitamins
    b. Cholesterol
    c. Saturated fat
    d. Total fat

Short Answer

1. What makes a protein able to perform its functions in the body?

2. During which points of life is the greatest amount of protein needed for building new tissue rapidly?

3. What does protein transport throughout the body, and how does this transport take place?

4. How can vegetarians eat a nutritionally adequate diet?

5. What is characterized by severe insufficiency of kcalories and protein, lack of fat stores, and muscle wasting?

6. What is the relationship between antigens and antibodies?

7. How do hormones move around the body?

8. How are proteins like carbohydrates and fats? How do they differ?

STUDENT WORKSHEET 5-1

TOPIC: Acceptable Macronutrient Distribution Ranges (AMDR)

1. Use a nutrient analysis program (your instructor will tell you which one to use) to analyze all the foods and beverages you had during one recent day.

2. Using your computerized output, write down in the first column how many kilocalories and grams of carbohydrate, fat, and protein you ate.

3. Multiply the number of grams of each macronutrient by the number of kcalories/gram given in the second column to come up with the number of kcalories from each macronutrient. Enter this number in the third column.

4. Now you can determine what percentage of your total kcalories came from each macronutrient. For example, if you ate 1200 kcalories from carbohydrate and your total kcalories for the day was 2400 kcalories, then divide 1200 by 2400 and multiply by 100 to get a percentage.

$$\frac{\text{1200 kcalories from carbohydrate}}{\text{2400 total kcalories}} \times 100 = 50\% \text{ of total kcalories from carbohydrate}$$

5. Finally compare your results to the AMDR. Did you eat within the limits of each AMDR? Chances are likely you did since the range within the AMDR is quite broad.

Total Kcalories:			AMDR
Carbohydrate Grams:	× 4 kcal/gram =	Kcalories from carb.	45-65%
Fat Grams:	× 9 kcal/gram =	Kcalories from fat	20-35%
Protein Grams:	× 4 kcal/gram =	Kcalories from protein	10-35%

STUDENT WORKSHEET 5-2

TOPIC: Comparing Protein Intake to the RDA

1. Tom is 5'11" tall and weighs 190 pounds. What is his RDA for protein?

2. Here are the foods that Tom ate yesterday. Using Appendix A in your text or another source, find out the amount of protein in each item he ate. Then total up the number of grams of protein.

Foods and Beverages (including portion size)	Grams of Protein
Breakfast	
1 1/2 cups corn flakes	
1 cup milk	
1 cup orange juice	
AM Snack	
3 1/2-inch bagel	
2 tablespoons cream cheese	
Lunch	
Double cheeseburger	
Medium Coke	
Medium French fries	
PM Snack	
Apple – medium	
Dinner	
Fried chicken, ¼ breast	
1 cup mashed potatoes	
1 cup canned corn	
1 cup milk	
Snack	
1 cup vanilla ice cream	
Total Protein:	

3. Compare Tom's protein intake to his RDA. Which foods contributed the most protein? Which were lowest in protein? Do you think you take in too much protein, just enough, or too little?

CHAPTER 6 VITAMINS

LEARNING OBJECTIVES

Upon completion of the chapter, the student should be able to:

1. State the general characteristics of vitamins
2. Identify the functions and food sources of each of the 13 vitamins
3. Identify which vitamins are most likely to be deficient in the American diet and which vitamins are most toxic
4. List two health benefits of eating a diet rich in fruits and vegetables
4. Discuss the use of fruits and vegetables on the menu
5. Describe ways to conserve vitamins when handling and cooking fruits and vegetables
6. Give examples of functional foods and discuss their role in the diet
7. Define phytochemicals and give examples of foods in which they are found

CHAPTER OUTLINE

1. **What are vitamins?**

   Vitamins are organic nutrients found in foods that are essential in small quantities for growth and good health. Vitamins are similar because they are made of the same elements—carbon, hydrogen, oxygen, and sometimes nitrogen or cobalt. They differ in the arrangement of their elements and the functions they perform in the body.

   Basics on vitamins:
   1. Very small amounts are needed by the body and very small amounts are in foods.
   2. The roles they play in the body are very important.
   3. Most vitamins are obtained through food. Some are made by bacteria in the intestine and one is made in the skin.
   4. There is no perfect food that contains all the vitamins in the right amount.
   5. Vitamins do not contain kcalories, but they are involved in extracting energy from the macronutrients.
   6. Some vitamins in foods are precursors.

   Distinguish between fat-soluble and water-soluble vitamins. The fat-soluble vitamins, as well as Vitamins B6 and B12, can be stored in the body.

2. **Discuss fat-soluble vitamins: A, D, E, and K**

   Fat-soluble vitamins are generally found in foods containing fats and are stored in the body either in the liver or in adipose tissue until needed. Excessive intake of A or D causes them to be stored and can be undesirable. Vitamin D, when taken in excess, is the most toxic of all the vitamins.

   Fat-soluble vitamins are absorbed and transported around the body like other fats. If anything interferes with fat absorption, these vitamins may not be absorbed.

   Dietary intake of vitamin E is low enough to be of concern for adults and children. Low intake of vitamin A is a concern for adults.

Vitamin	Recommended Intake	Functions	Sources
Vitamin A	*RDA*: Men: 900 micrograms RAE Women: 700 micrograms RAE *Upper Intake Level*: 3,000 micrograms/day of preformed vitamin A	Health of eye (especially cornea and retina), vision Epithelial cells that form skin and protective linings of lungs, gastrointestinal tract, urinary tract, and other organs Reproduction Growth and development Bone and teeth development Immune system function Antioxidant	*Preformed*: Liver, fortified milk, fortified cereals, eggs. *Beta carotene*: Dark green vegetables, deep orange fruits and vegetables.
Vitamin D	*AI:* 5 micrograms cholecalciferol (31 to 50 years old) 10 micrograms cholecalciferol (51 to 70+ years old) *Upper Intake Level:* 50 micrograms cholecalciferol	Maintenance of blood calcium levels so calcium can build bones and teeth. Bone growth	Sunshine, fortified milk, fortified cereals, fatty fish, fortified butter and margarine.
Vitamin E	*RDA:* 15 mg alpha-tocopherol *Upper Intake Level:* 1,000 mg alpha-tocopherol (synthetic forms from supplements and/or fortified foods)	Antioxidant – especially helps red blood cells and cells in lungs and brains. Protection of vitamin A from oxidation	Vegetable oils, margarine, shortening, salad dressing, nuts, seeds, leafy green vegetables, whole grains, fortified cereals.
Vitamin K	*AI*: Men: 120 micrograms Women: 90 micrograms	Blood clotting Healthy bones	Green leafy vegetables, vegetable oils and margarines Made in intestine

Nutrition Science Focus:
Vitamin A: Without enough vitamin A, you can develop night blindness. Also the cornea of the eyes becomes cloudy and dries (called xerosis) and thickens, and this can result in permanent blindness (xerophthalmia).

Vitamin D: In its active form vitamin D acts more as a hormone than a vitamin. The active form of vitamin D travels through the bloodstream to increase calcium absorption in the intestine, decrease the amount of calcium excreted by the kidneys, and pulls calcium out of the bones. Research also suggests that vitamin D may help maintain a healthy immune system and help regulate cell growth and differentiation (the process that determines what a cell is to become).

Vitamin E: Vitamin E is of particular importance to cell membranes at the highest risk of oxidation —including cells in the lungs, red blood cells, and brain. Vitamin E even protects vitamin A from oxidation.

3. **Discuss water-soluble vitamins: vitamin C and B-complex**
   The B vitamins work in everybody cell where they function as coenzymes.

   The body stores only limited amounts of water-soluble vitamins (except vitamins B6 and B12). Therefore, these vitamins need to be taken in daily. Excessive supplementation of certain water-soluble vitamins can still cause toxic side effects.

   Dietary intake of vitamin C is low for many American adults.

   Choline is considered a conditionally essential nutrient, because when the diet contains no choline, the body can't make enough of it and liver damage can result.

Vitamin	Recommended Intake	Functions	Sources
Vitamin C	*RDA:* Men: 90 mg Women: 75 mg *Upper Intake Level:* 2,000 milligrams	Collagen formation Wound healing. Synthesis of some hormones	Citrus fruits, bell peppers, kiwi fruit, strawberries, tomatoes, broccoli, potatoes; fortified juices, drinks, and cereals
Thiamin	*RDA:* Men: 1.2 mg Women: 1.1 mg *Upper Intake Level:* None	Part of coenzyme in energy metabolism Nerve function	Pork, dry beans, whole-grain and enriched/fortified breads and cereals, peanuts, acorn squash.
Riboflavin	*RDA:* Men: 1.3 mg Women: 1.1 mg *Upper Intake Level:* None	Part of coenzymes in energy metabolism Formation of vitamin B6 coenzyme and niacin	Milk, milk products, organ meats, whole-grain and enriched/fortified breads and cereals, eggs, some meats

Niacin	*RDA:* Men: 16 mg niacin equivalent Women: 14 mg niacin equivalent *Upper Intake Level:* 35 mg niacin equivalents (synthetic forms from supplements and/or fortified foods)	Part of coenzymes in energy metabolism	Meat, poultry, fish, organ meats, wholegrain and enriched/fortified breads and cereals, peanut butter, milk, eggs.
Vitamin B6	*RDA:* 1.3 mg *Upper Intake Level:* 100 mg	Part of coenzyme involved in carbohydrate, fat, and especially protein metabolism. Synthesis of hemoglobin (in red blood cells) and some neurotransmitters Important for immune system	Meat, poultry, fish, potatoes, fruits such as bananas, some leafy green vegetables, fortified cereals.
Folate	*RDA:* 400 micrograms dietary folate equivalent *Upper Intake Level:* 1,000 micrograms dietary folate equivalents (synthetic forms from supplements and/or fortified foods)	Part of coenzyme required to make DNA and new cells. Amino acid metabolism	Fortified cereals, green leafy vegetables, legumes, orange, fortified breads.
Vitamin B12	*RDA:* 2.4 micrograms *Upper Intake Level:* None	Part of coenzyme that makes new cells and DNA Conversion of folate into active coenzyme form Normal functioning of nervous system Healthy bones	Animal foods such meat, poultry, fish, shellfish, eggs, milk, and milk products
Pantothenic Acid	*AI:* 5 mg *Upper Intake Level:* None	Part of coenzyme in energy metabolism	Widespread Fortified cereals, beef, poultry, mushrooms, potatoes, tomatoes

Biotin	*AI:* 30 micrograms *Upper Intake Level:* None	Part of coenzyme involved in amino acid metabolism, and synthesis of fat and glycogen.	Widespread Egg yolks Made in intestine
Choline (Conditionally essential)	*AI:* Men: 550 mg Women: 425 mg *Upper Intake Level:* 3500 mg	Synthesis of neurotransmitter Synthesis of lecithin (a phospholipid) found in cell membranes.	Widespread. Milk, eggs, peanuts

Nutrition Science Focus:
Vitamin C: Fig 6-5 shows how vitamin C acts as an antioxidant. Ascorbic acid donates two of its hydrogens to the free radicals to neutralize them and prevent them from damaging DNA or other substances. This substances can convert back to ascorbic acid. One additional function of vitamin C is that it is needed to convert certain amino acids into the neurotransmitters serotonin and norepinephrine.

Thiamin, Riboflavin, and Niacin: Thiamin also plays a vital role in the normal functioning of the nerves. Riboflavin is part of coenzymes that help form the vitamin B6 coenzyme and make niacin in the body from tryptophan.

Vitamin B6: Vitamin B6 is also used to break down glycogen to glucose to keep your blood sugar level steady during the night. Vitamin B6 is also needed for the synthesis of neurotransmitters such as serotonin and dopamine.

Folate and Vitamin B12: Folate and vitamin B12 are both involved with making DNA and new cells. They need each other to become active in the body.

4. **Discuss fruits and vegetables**
   Nutrition: Fruits and vegetables are low in kcalories, low in fat (or no fat), no cholesterol, good sources of fiber, low in sodium, and excellent sources of vitamins and minerals.

5. **Culinary Science**
   Once picked, the quality of fruits and vegetables starts to deteriorate. To slow down this process, you need to keep produce cold and moist. Once produce cells lose water, the cells shrink and the produce looks limp or wilted.

   Fruits and vegetables are mostly carbohydrates and they cook evenly and become soft and tender, although you have to be careful about getting the right texture. The fiber in fruits and vegetables can be made firmer by adding sugar, as an acid such as lemon juice, to cooking vegetables.

The longer a vegetable is cooked, the more flavor is lost into the cooking liquid and into the air through evaporation. To decrease flavor loss, it is best to cook in just enough water to cover and cook for as short a time as possible. The best way to cook vegetables is to steam them—this method helps retain flavor and nutrients. Microwaving is also good.

Finally, you need to control changes in color as the vegetable cooks. Prolonged cooking of green vegetables causes them to turn drab olive green. The pigments in yellow and orange vegetables are very heat stable. Overcooking turns white vegetables a dull gray. Red vegetables turn brighter red in acids and blue or blue-green in alkali solutions such as baking soda.

6. **Food Facts: Functional Foods: Superfoods**
   Functional foods are foods found in a usual diet that have biologically active components, such as lycopene found in tomato products, which may provide health benefits beyond basic nutrition. Functional foods also include foods that have added vitamins, minerals, phytochemicals, and/or herbs. Many of the health-promoting ingredients in functional foods are phytochemicals—bioactive compounds found in plants that are linked to decreased risk of chronic diseases.

   Examples of functional foods which you might want to think of as superfoods because they are good for your health: beans, nuts, cocoa, tea, spinach, plant stanols and sterols.

   The use of probiotics and prebiotics are promising areas for functional foods. Probiotics refers to adding health-promoting bacteria to foods such as yogurt. A prebiotic is a nondigestible food ingredient such as fiber which stimulate the growth and activity of health-promoting bacteria in the colon.

   When manufacturers add herbs to foods, they don't have to disclose how much is added. Questions to consider:
   - Is it safe?
   - Do herbs, or other added ingredients, really work?
   - Will making tea with St. John's Wort really help your depression?
   - Will beta-carotene in a food it normally doesn't appear in still work like the beta-carotene in a carrot?
   - Will any of the added ingredients interact with a medication or dietary supplement you take?
   - Is the fortified food a healthy one, or is it a fortified candy bar?

7. **Hot Topic: Phytochemicals**
   Phytochemicals are substances such as beta-carotene, found largely in fruits and vegetables, which seem to be helpful in preventing cancer and/or heart disease. Examples: lycopene in tomatoes, phenols in tea, sulforaphane in broccoli. You can get in phytochemicals by eating 5 to 9 servings a day of fruits and vegetables.

NUTRITION WEB EXPLORER

Food and Drug Administration: www.cfsan.fda.gov/~dms/supplmnt.html
This is a special FDA site related to supplements. Click on "Questions and Answers" and find out what a dietary supplement is and where you can get information about a specific dietary supplement.

Centers for Disease Control and Prevention: www.fruitsandveggiesmatter.gov
On this home page for the Fruits and Vegetables campaign, do the exercise "How Many Fruits and Vegetables Do You Need?"
Then click on "What Counts as a Cup?" How many grapes, baby carrots, raisins, or grapefruit count as 1 cup?
Next, click on recipes. Select one recipe and do a nutrient analysis of it using ***iProfile***. Which vitamin is highest in your recipe? Compare to others in your class.

CHAPTER REVIEW QUIZ

Key Terms: Matching

1. Precursors
2. Retinol
3. Carotenoid
4. Beta-carotene
5. Xerosis
6. Xerophthalmia
7. Antioxidant
8. Free radical
9. Cholecalciferol
10. Rickets
11. Osteomalacia
12. Alpha-tocopherol
13. Scurvy
14. Collagen
15. Tryptophan
16. Intrinsic factor

a. A disease of vitamin D deficiency in adults in which the leg and spinal bones soften and may bend
b. A protein-like substance secreted by stomach cells that is necessary for the absorption of vitamin B12
c. A precursor of vitamin A that functions as an antioxidant in the body; the most abundant carotenoid
d. The form of vitamin D found in animal foods
e. Forms of vitamins that the body changes chemically to active vitamin forms
f. An unstable compound that reacts quickly with other molecules in the body
g. A form of vitamin A found in animal foods
h. A childhood disease in which bones do not grow normally, resulting in bowed legs and knock knees
i. A class of pigments that contribute a red, orange, or yellow color to fruits and vegetables
j. An amino acid present in protein foods that can be converted to niacin in the body
k. The most active form of vitamin E in humans; also a powerful antioxidant
l. Hardening and thickening of the cornea that can lead to blindness; usually caused by a deficiency of vitamin A
m. A condition in which the cornea of the eye becomes dry and cloudy; often due to a deficiency of vitamin A
n. A vitamin C deficiency disease marked by bleeding gums, weakness, loose teeth, and broken capillaries under the skin
o. A compound that combines with oxygen to prevent oxygen from oxidizing or destroying important substances
p. The most abundant protein in the body

Multiple Choice

1. Which vitamins are readily excreted by the body?
   a. Vitamin A
   b. Vitamin B1
   c. Vitamin B6
   d. Vitamin B12

2. This vitamin, which is the most abundant carotenoid, is split in the intestine and liver to make retinol.
   a. Retinoid
   b. Retinal
   c. Beta-carotene
   d. Niacin

3. This vitamin is needed to produce and maintain the epithelial cells that form the protective linings of the lungs, GI tract, and urinary tract.
   a. Vitamin C
   b. Vitamin D
   c. Vitamin A
   d. Vitamin E

4. This vitamin, when taken in excess of the AI, is the most toxic of all vitamins.
   a. Vitamin K
   b. Vitamin A
   c. Vitamin E
   d. Vitamin D

5. Tryptophan, an amino acid present in some foods, can be converted to which vitamin in the body?
   a. Folic acid
   b. Niacin
   c. Riboflavin
   d. Thiamin

6. Pregnancy and nursing, growth, burns, surgery, fractures, and cigarette smoking require additional:
   a. Vitamin B6
   b. Vitamin C
   c. Vitamin B12
   d. Beta-carotene

7. Good sources of folate include all the following, except:
    a. Salmon
    b. Legumes
    c. Leafy green vegetables
    d. Orange juice

8. ___________ functions primarily as part of coenzymes.
    a. Vitamin A
    b. Vitamin B
    c. Vitamin C
    d. Vitamin D

9. All of the following vitamins are antioxidants except:
    a. Vitamin C
    b. Vitamin K
    c. Vitamin E
    d. Vitamin A

10. Dietary intake of this vitamin is a concern for both children and adults:
    a. Choline
    b. Vitamin E
    c. Vitamin D
    d. Pantothenic acid

True/False

1. Some vitamins in foods are precursors to the active form of the vitamin rather than active vitamins themselves.
    a. True   b. False

2. All B vitamins must be obtained through foods because the body does not produce enough of them alone.
    a. True   b. False

3. Permanent blindness resulting from the hardening and thickening of the cornea and caused by vitamin A deficiency is called Xerosis.
    a. True   b. False

4. Cholecalciferol is the inactive form of vitamin D.
    a. True   b. False

5. In its active form vitamin D functions more as a hormone than as a vitamin.
    a. True   b. False

6. Vitamin A protects vitamin E from oxidation.
    a. True   b. False

7. Toxicity is a problem with vitamin K consumption if the Tolerable Upper Intake Level has been exceeded.
   a. True   b. False

8. Pernicious anemia develops when B12 is not properly absorbed in the body.
   a. True   b. False

9. Pork is an excellent source of riboflavin.
   a. True   b. False

10. Collagen is a muscle that provides support to bones, teeth, skin, and cartilage.
   a. True   b. False

Short Answer

1. Which types of vitamins generally occur in foods containing fat and are usually stored in the liver or adipose tissue until needed?

   ADEK (fat soluble

2. Toxicity of which vitamin can cause bleeding problems?

   E

3. Which vitamin differs from the others in that it can be made in the body?

   D

4. What is the defect caused by folic acid deficiency where most of the brain is missing?

   Anencephaly

5. Which vitamin is considered a conditionally essential nutrient because without it liver damage can occur?

   Choline

STUDENT WORKSHEET 6-1

TOPIC: Nutrient Analysis of Three-Day Diet

1. Use ***iProfile*** to analyze all the foods and beverages you had during a three day period, preferably including two weekdays and one weekend day. Use the attached Daily Food Record to record your food intake.

2. Using your computerized output, write down below what percent of the dietary recommendations you had for each vitamin.

Vitamins	% Recommendation
Vitamin A	
Vitamin D	
Vitamin E	
Vitamin K	
Vitamin C	
Thiamin	
Riboflavin	
Niacin	
Vitamin B6	
Folate	
Vitamin B12	
Pantothenic Acid	
Biotin	
Choline	

3. Circle the vitamins in which you are deficient. List foods below that you could eat to increase your consumption for each of these vitamins.

DAILY FOOD RECORD

Name _________________________

Day and Date ___________________

Name of Food	Portion Size

STUDENT WORKSHEET 6-2

TOPIC: Vitamin Salad Bar

You are to set up a salad bar using any foods you like as long as you have a good source of each of the 13 vitamins. Fill in the circles below as though this is your layout for the salad bar. In the spaces below, write down the name of the food and which vitamin(s) it is rich in.

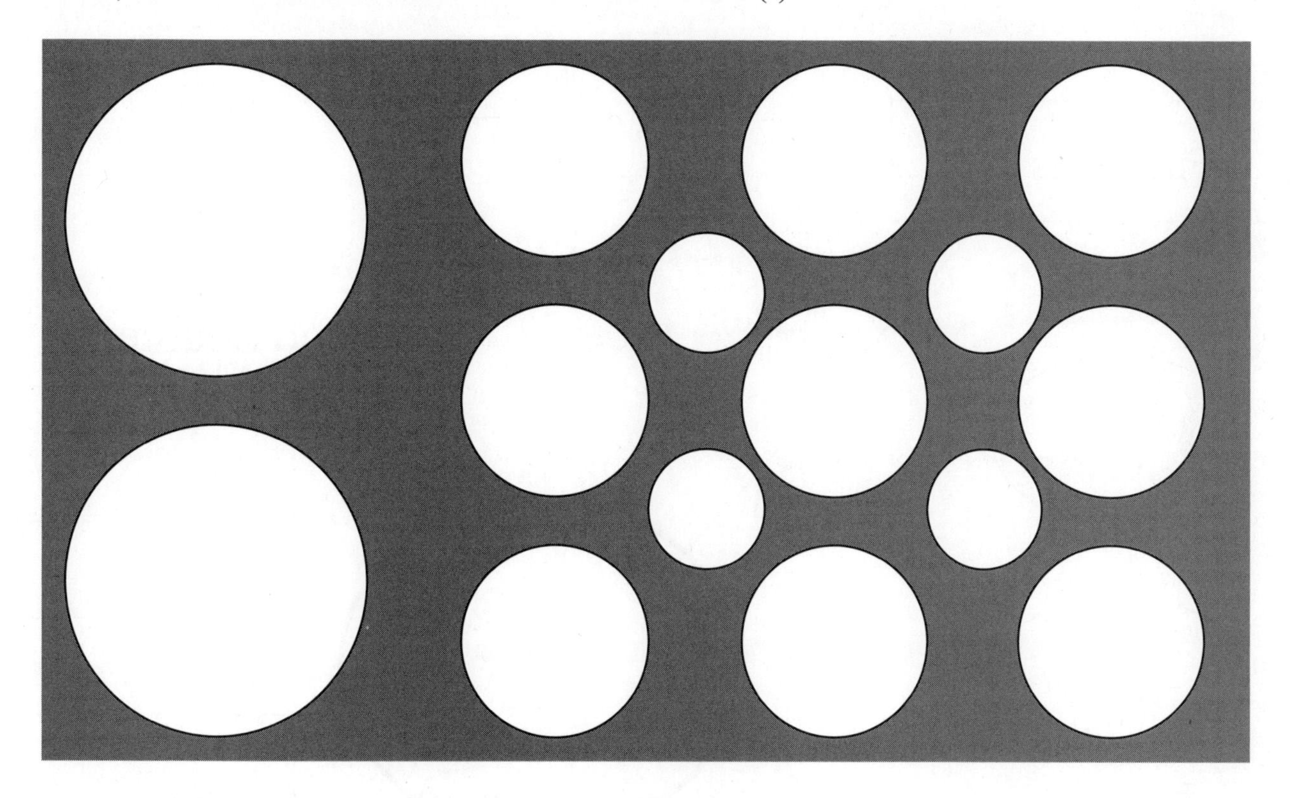

	Ingredients	Vitamins
Large Bowl		
Large Bowl		
Medium Bowl		
Medium Bowl		
Medium Bowl		
Medium Bowl		
Medium Bowl		
Medium Bowl		
Medium Bowl		
Medium Bowl		
Medium Bowl		
Small Bowl		
Small Bowl		
Small Bowl		
Small Bowl		

STUDENT WORKSHEET 6-3

TOPIC: Concept Mapping

Vitamin Name: ______________________________

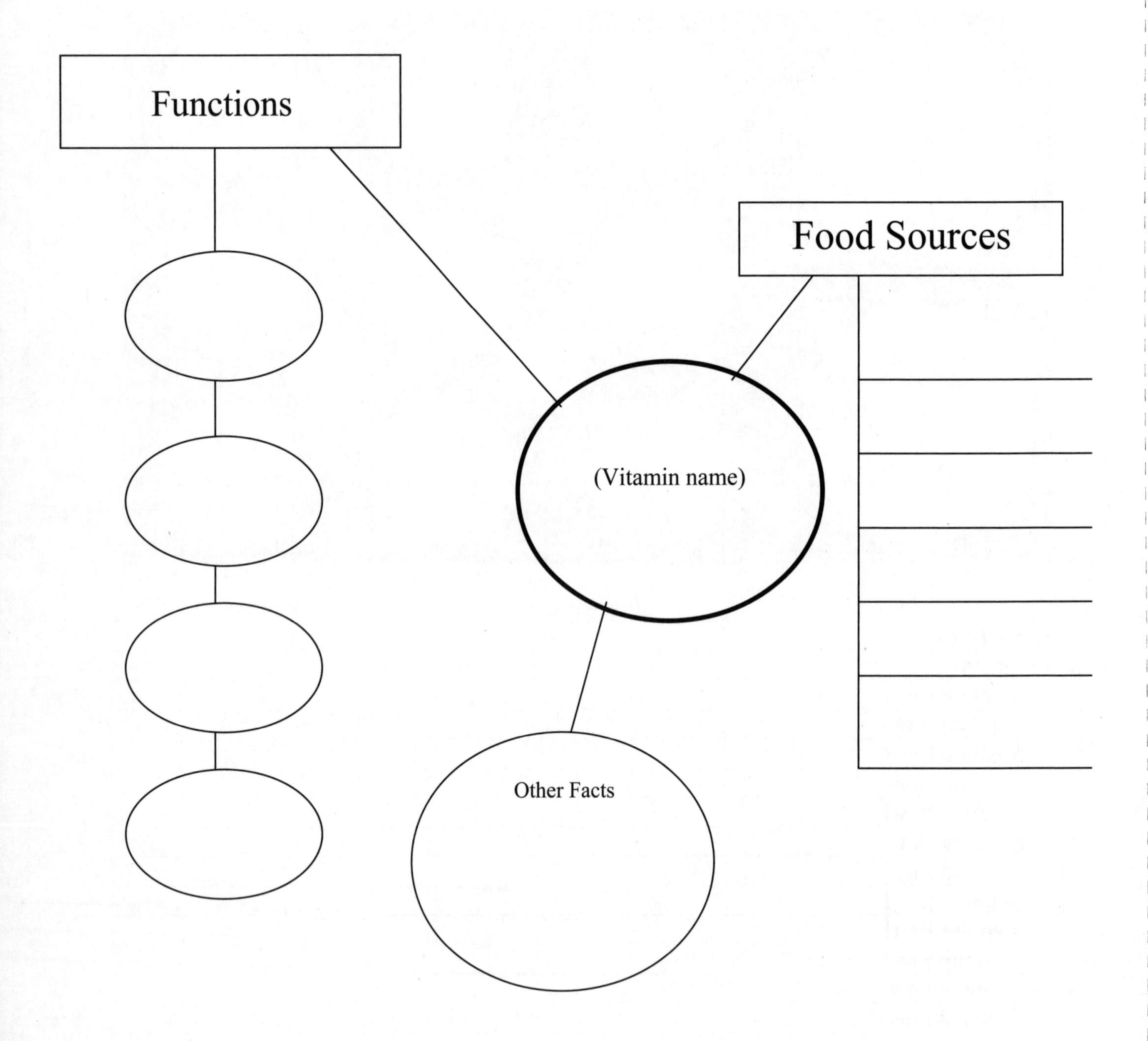

CHAPTER 7 WATER AND MINERALS

LEARNING OBJECTIVES

Upon completion of the chapter, the student should be able to:

1. State the general characteristics of minerals
2. Identify the percentage of body weight made up of water
3. List the functions of water in the body
4. Identify the functions and food sources of the major minerals and the trace minerals
5. Identify which minerals are most likely to be deficient in the American diet
6. Discuss the nutrition of nuts and seeds and how to use them on the menu
7. Distinguish between different types of bottled waters
8. Explain how dietary supplements are regulated and labeled
9. Identify instances when supplements may be necessary

CHAPTER OUTLINE

1. **What are minerals?**

   You need only small amounts of minerals but they do enormously important jobs in your body.

   Major minerals are needed in relatively large amounts in the diet. Trace minerals are needed in smaller amounts—less than 100 milligrams daily.

   The percentage of minerals that is absorbed varies tremendously (bioavailability). Minerals in animal foods tend to be absorbed better than those in plant foods.

   Minerals are not destroyed in food storage or preparation, except they can be lost in cooking liquids.

   Minerals can be toxic when consumed in excess, and may interfere with the absorption and metabolism of other minerals.

2. **Water**

   The average adult's body weight is 50–60 percent water.

   Functions of water
   - *Medium for many metabolic activities and also participates in some metabolic reactions.*
   - *Carries nutrients to the cells and carries away waste materials to the kidneys and out of the body in urine.*
   - *Needed in each step of the process of converting food into energy and tissue.*
   - *Maintains blood volume in your body.*
   - *Maintain normal body temperature.*
   - *Important part of body lubricants, such as cushioning joints and internal organs.*

   AI for **total** water: Men: 3.7 liters/day
   Women: 2.7 liters/day
   Hydration needs can be met through drinking juice, milk, coffee, tea, soda, etc. and by eating foods such as fruits and vegetables that are high in water.

A number of mechanisms, including thirst, operate to keep body water content within narrow limits. The body gets rid of the water it doesn't need through the kidneys and skin and, to a lesser degree, from the lungs and gastrointestinal tract.

Bottled water is different from tap water in that it has more consistent quality and taste. The taste of water has to do with the way it is treated and the quality of its sources, including its natural mineral content. One of the key taste differences between tape water and bottled water is how the water is disinfected. Tap water is often disinfected with chlorine while bottled water is often disinfected with ozone. While bottled water originates from protected sources such as springs, tap water comes mostly from lakes and rivers. Bottled water is regulated by the FDA.

3. **Discuss major minerals**

Discuss functions and sources

Calcium	*AI:* 1,000 mg *Upper Intake Level:* 2500 mg	Building bones and teeth Blood clotting Muscle contraction Nerve transmission May lower blood pressure	Milk and many milk products, calcium-fortified foods, tofu made with calcium carbonate, several greens (broccoli, collards), legumes, whole wheat bread
Phosphorus	*RDA:* 700 mg *Upper Intake Level:* 4,000 mg	Building bones and teeth Energy metabolism Part of DNA Buffer of acids and bases	Milk and milk products, meat, poultry, fish, eggs, legumes
Magnesium	*RDA*: Men: 420 mg Women: 310 mg *Upper Intake Level*: 350 mg	Energy metabolism Bones and teeth Muscle contraction Nerve transmission Immune system	Nuts, seeds, legumes, green leafy vegetables, potatoes, whole-grain breads and cereals, seafood
Sodium	*AI*: 1500 mg *Upper Intake Level:* 2300 mg	Water balance Acid-base balance Muscle contraction Nerve transmission	Salt, processed foods such as luncheon meats, salted snacks, canned foods), soy sauce
Potassium	*AI:* 4700 mg *Upper Intake Level*: None	Water balance Acid-base balance Muscle relaxation Nerve transmission	Fruits and vegetables (winter squash, potatoes, oranges, grapefruits, banana), milk, yogurt, legumes, meat

Sodium Chloride = Salt
Sodium - metal
Chloride - gas

Chloride	*AI*: 2300 mg *Upper Intake Level:* 3600 mg	Water balance Acid-base balance Part of hydrochloric acid in stomach	Salt
Sulfur	None	Part of protein, thiamin, and biotin	Protein foods

Nutrition Science Focus:

Calcium and Phosphorus: Bones undergo continuous remodeling with bone formation and bone resorption occurring constantly. Bones stop increasing in density after age 30. Vitamin D and 2 hormones, parathyroid and calcitonin, work to keep blood calcium at just the right level. Phosphorus activates many enzymes when a phosphate group is attached. Phosphorus is also present in phospholipids.

Magnesium: Magnesium inhibits contraction and promotes relaxation of muscles. Magnesium appears to be a heart-healthy nutrient.

Electrolytes: About two-thirds of your body's fluids are inside your cells (intercellular fluids). Potassium is almost completely found inside the cell. About one-third of your body's fluids are extracellular fluids—fluids found outside of your cells. This includes fluids in your blood vessels and in the interstitial spaces (between the cells). Sodium and chloride are mostly in the extracellular fluids. The distribution and balance of the body's fluids are essential to the normal functioning of the body. The kidney is one organ that is involved in maintaining this balance.

4. **Discuss trace minerals**
   Many trace minerals are toxic at levels only several times higher than recommendations. Also trace minerals are highly interactive with each other.

Iron	*RDA:* Men: 8 mg Women: 18 mg *Upper Intake Level*: 45 mg	Component of hemoglobin and myoglobin Energy metabolism	Beef, poultry, fish, enriched and fortified breads and cereals, legumes, green leafy vegetables, eggs

ADDITIONAL NOTES:

- Only about 15% of dietary iron is absorbed—heme iron (in animal foods) is absorbed and used twice as readily as iron in plant foods (nonheme iron).
- Vitamin C, meat, poultry, and fish all increase iron absorption when eaten along with the iron-containing food.
- Iron deficiency is the most common nutritional deficiency. If severe enough, it results in iron-deficiency anemia (size and number of RBC are reduced). Half of all pregnant women develop iron-deficiency anemia which can create problems for the baby.
- Iron overload is a common genetic disease in which individuals absorb about twice as much iron from food and supplements as other people.

Zinc	*RDA:* Men: 11 mg Women: 8 mg *Upper Intake Level:* 40 mg	Wound healing Bone formation DNA synthesis Protein, carbohydrate, and fat metabolism Development of sexual organs Taste perception	Shellfish, meat, poultry, legumes, dairy foods, whole grains, fortified cereals

ADDITIONAL NOTES:

- Only about 40% of zinc we eat is absorbed. Phytates can decrease absorption.
- Deficiencies are more likely to show up in the young, the elderly, and pregnant women.
- Zinc also is involved in general tissue growth and maintenance, vitamin A activity, protection of cell membranes from free-radical attacks, and participates in storing and releasing insulin.

Iodine	*RDA*: 150 micrograms *Upper Intake Level*: 1100 micrograms	Part of thyroid hormones Normal metabolic rate	Iodized salt, saltwater fish, grains and vegetables grown in iodine-rich soil

ADDITIONAL NOTES:

- A deficiency can cause hypothyroidism (less thyroid hormone is made leading to low metabolic weight) and simple goiter (enlarged thyroid gland).
- An iodine deficiency during pregnancy can harm the fetus and cause cretinism (condition of mental and physical retardation).

Selenium	*RDA*: 55 micrograms *Upper Intake Level:* 400 micrograms	Part of antioxidant enzymes Regulates thyroid hormone	Grains and vegetables grown in selenium-rich soil, seafood, meat
Fluoride	*AI*: Men: 4 mg Women: 3 mg *Upper Intake Level:* 10 mg	Strong teeth and bones	Fluoridated water, fish, tea

ADDITIONAL NOTES:

- Too much fluoride can lead to fluorosis (teeth become mottled and disordered).
- Fluoride taken internally, whether in drinking water or dietary supplements, can strengthen babies' and children's developing teeth to resist decay.

Chromium	*AI:* Men: 35 micrograms Women: 25 micrograms *Upper Intake Level:* None	Works with insulin	Unprocessed foods such as whole grains, broccoli, nuts, egg yolks, green beans
Copper	*RDA:*	Works with iron to form	Liver, seafood, nuts,

	900 micrograms *Upper Intake Level:* 10,000 micrograms	hemoglobin Synthesis of collagen Energy metabolism	seeds, beans

ADDITIONAL NOTES:

- Deficiency is rare.
- Single doses of copper only 4 times the recommended level can cause vomiting and nervous system disorders.

Manganese	*AI*: Men: 2.3 m. Women: 1.8 mg *Upper Intake Level:* 11 mg	Form bone Metabolism of carbohydrate, fat, and protein	Whole grains, dried fruits, nuts, leafy vegetables
Molybdenum	*RDA*: 45 micrograms *Upper Intake Level:* 2,000 micrograms	Cofactor for several enzymes	Legumes, meat, whole grains, nuts

Possible candidates for nutrient status include arsenic, boron, nickel, silicon, and vanadium. Based on adverse effects noted in animal studies, UL have been set for boron, nickel, and vanadium.

5. **Nuts and Seeds**
Nuts and seeds pack many minerals and vitamins, along with fiber and protein, in their small sizes. Nuts especially also contain quite a bit of fat, but the fat is mostly monounsaturated.

6. **Culinary Science**
Cooking nuts is a quick process that gives them more flavor and softens their chewy texture. As nuts cool after cooking, their texture becomes crispy.

7. **Food Facts: How to Retain Vitamins and Minerals From Purchasing to Serving**
Five factors are responsible for most nutrient loss: heat, exposure to the air and light, cooking in water, and baking soda. Tips to retain nutrients are discussed.

8. **Hot Topic: Dietary Supplements**
In addition to vitamins and minerals, dietary supplements may include herbs, botanicals, and other plant-derived substances, as well as amino acids, concentrates, metabolites, constituents and extracts of these substances.

The 1994 Dietary Supplement Health and Education Act (DSHEA) requires manufacturers to include the words "Dietary Supplement" on labels as well as a "Supplement Facts" panel.

Dietary supplements are NOT drugs or replacements for conventional diets.

Under DSHEA and previous food labeling laws, supplements may use, as appropriate, the following types of claims: nutrient-content claims ("high in calcium"), disease claims ("folate is important to reduce neural tube defects during pregnancy"), and nutrition-support claims, which include structure-function claims ("calcium builds strong bones").

Discuss ways that consumers can protect themselves when buying supplements, such as looking for USP notation, limiting intake to 100% of the DRI, avoiding substances that are not known nutrients, and considering nationally known manufacturers.

Situations when supplements may be needed include: women in their childbearing years, pregnant/lactating women, people with known nutrient deficiencies, elderly people who are eating poorly, drug addicts or alcoholics, people eating less than 1,200 kcalories a day, people on certain medications or with certain diseases.

NUTRITION WEB EXPLORER

Oregon Dairy Council: www.oregondairycouncil.org/calcium_checkup/
Click on "Calcium: Are You Getting Enough?" and see how much calcium is in your diet..

American Heart Association: www.americanheart.org/presenter.jhtml?identifier=582
Take the "Sodium Intake Quiz" to determine how much sodium you take in each day.

Medline Plus Health Information on Supplements:
http://www.nlm.nih.gov/medlineplus/minerals.html
Click on a topic under "Latest news" and write a summary of what you read.

National Center for Complementary and Alternative Medicine:
www.nccam.nih.gov

The National Center for Complementary and Alternative Medicine is dedicated to exploring complementary and alternative healing practices in the context of rigorous science. Read "What is CAM" to learn about this new area.

Then write a summary paragraph.

Office of Dietary Supplements, National Institutes of Health:
http://dietary-supplements.info.nih.gov
Click on "Health Information" and then "The Savvy Supplement User." Write down any tips that are useful for you or someone you know, such as a grandmother, who is taking supplements.

Iron Overload Diseases Association: www.ironoverload.org
Click on "Diet for Hemochromatosis, Iron Overload, and Anemia." What foods should be avoided if you have iron overload?

CHAPTER REVIEW QUIZ

Key Terms: Matching

1. Major minerals
2. Trace minerals
3. Hydroxy-apatite
4. Oxalic acid
5. Phytic acid
6. Electrolytes
7. Ion
8. Hemoglobin
9. Myoglobin
10. Heme iron
11. Nonheme iron
12. Hemochroma-tosis
13. Thyroid gland
14. Hypothyroid-ism
15. Simple goiter
16. Cretinism

a. Thyroid enlargement caused by inadequate dietary intake of iodine.
b. A form of iron found in all plant sources of iron and also as part of the iron in animal food sources.
c. An atom or group of atoms carrying a positive or negative electric charge.
d. Minerals needed in smaller amounts in the diet-less than 100 milligrams daily.
e. A condition in which there is less production of thyroid hormones.
f. A binder found in wheat bran and whole grains that can decrease the absorption of certain nutrients, such as calcium and iron.
g. A gland found on either side of the trachea that regulates the level of metabolism.
h. Mental and physical retardation during fetal and later development caused by iodine deficiency during pregnancy.
i. The main structural component of bone composed mostly of calcium phosphate crystals.
j. A common genetic disease in which individuals absorb about twice as much iron from their food and supplements as other people do.
k. A protein in red blood cells that carries oxygen to the body's cells.
l. The predominant form of iron in animal foods; it is absorbed and used more readily than iron in plant foods.
m. Minerals needed in relatively large amounts in the diet-over 100 milligrams daily.
n. A muscle protein that stores and carries oxygen that the muscles will use to contract.
o. An organic acid found in spinach and other leafy green vegetables that can decrease the absorption of certain minerals, such as calcium.
p. Chemical elements or compounds that ionize in solution and can carry an electric current.

Multiple Choice

1. All of the following are considered trace minerals except:
   a. Calcium
   b. Zinc
   c. Iron
   d. Fluoride

2. Water is responsible for all of the following in the body except:
   a. Participating in metabolic reactions
   b. Carrying nutrients away from cells
   c. Carrying away waste materials to the kidneys
   d. Carrying waste from the body in urine

3. Which of the seven major minerals is the least likely to cause concern due to low consumption?
   a. Phosphorus
   b. Potassium
   c. Calcium
   d. Magnesium

4. Magnesium is essential for all of the following except:
   a. Building bones and maintaining teeth
   b. Stabilizing fluid levels in the body
   c. Muscle relaxation
   d. Energy metabolism

5. Electrolytes are important because:
   a. They maintain water balance
   b. They act as buffers to neutralize various acids and bases in the body
   c. They act together to lower blood pressure
   d. Both a and b
   e. All of the above

6. If a woman has a severe iodine deficiency during pregnancy, the fetus can have a condition of mental and physical retardation known as:
   a. Hypothyroidism
   b. Hyperthyroidism
   c. Cretinism
   d. Simple goiter

7. This essential mineral is also part of an enzyme that activates thyroid hormones:
   a. Selenium
   b. Fluoride
   c. Chromium
   d. Potassium

8. This mineral works with insulin to transfer glucose from the bloodstream into the body's cells.
   a. Sodium
   b. Nitrogen
   c. Chromium
   d. Zinc

9. The richest source of selenium is:
   a. Turkey breast
   b. Walnuts
   c. Cod
   d. Brazil nuts

10. Which vitamin plays an important role in calcium absorption and bone health?
    a. Vitamin A
    b. Vitamin E
    c. Vitamin D
    d. Vitamin C

True/False

1. Unlike vitamins, minerals are organic elements that are not destroyed in food storage or preparation.
   a. True   b. False

2. The average adult's body weight is generally 50 to 60 percent water.
   a. True   b. False

3. One of water's functions is to maintain blood volume in the body.
   a. True   b. False

4. Oxalic acid, a binder that prevents some calcium from being absorbed, is found in wheat bran and whole grains.
   a. True   b. False

5. A diet low in potassium blunts the effects of salt on blood pressure and see to lower blood pressure overall.
   a. True   b. False

6. There is no evidence of chronic excess intake of potassium in healthy individuals so no Tolerable Upper Intake Level has been set.
   a. True   b. False

7. Calcium is a part of Myoglobin, a muscle protein that stores and carries oxygen that muscles use to contract.
   a. True
   b. False

8. People who exercise may require vitamin D supplementation.
   a. True
   b. False

Short Answer

1. Which minerals form part of hydroxyapatite which gives bone its strength?

   Calcium + phosphorous

2. Sodium, potassium, and chloride are collectively referred to as electrolytes.

3. What are four symptoms of iron-deficiency anemia?

   feeling tired + weak, decrease work performance, difficulty maintaining body temp, + decreased immune function

4. Which mineral, when taken in excess, causes copper deficiency?

   Zinc

5. In its most severe form, ___________ causes a distinct brownish mottling.

   fluorosis

STUDENT WORKSHEET 7-1

TOPIC: Nutrient Analysis of Three-Day Diet

1. Use ***iProfile*** to analyze all the foods and beverages you had during a three-day period, preferably including two weekdays and one weekend. Use the attached Daily Food Record to record your food intake.

2. Using your computerized output, write down below what percent of the dietary recommendations you had for each mineral.

Minerals	% Recommendation
Calcium	
Phosphorus	
Magnesium	
Sodium	
Potassium	
Iron	
Zinc	
Selenium	
Chromium	
Copper	
Manganese	
Molybdenum	

3. Circle the minerals in which you are deficient. List foods below that you could eat to increase your consumption for each of these minerals.

DAILY FOOD RECORD

Name ______________________

Day and Date _________________

Name of Food	Portion Size

STUDENT WORKSHEET 7-2

TOPIC: Mineral Salad Bar

You are to set up a salad bar using any foods you like as long as you have a good source of each of these minerals: calcium, magnesium, potassium, iron, zinc, selenium, chromium, copper, manganese, and molybdenum. Fill in the circles below as though this is your layout for the salad bar. In the spaces below, write down the name of the food and which mineral(s) it is rich in.

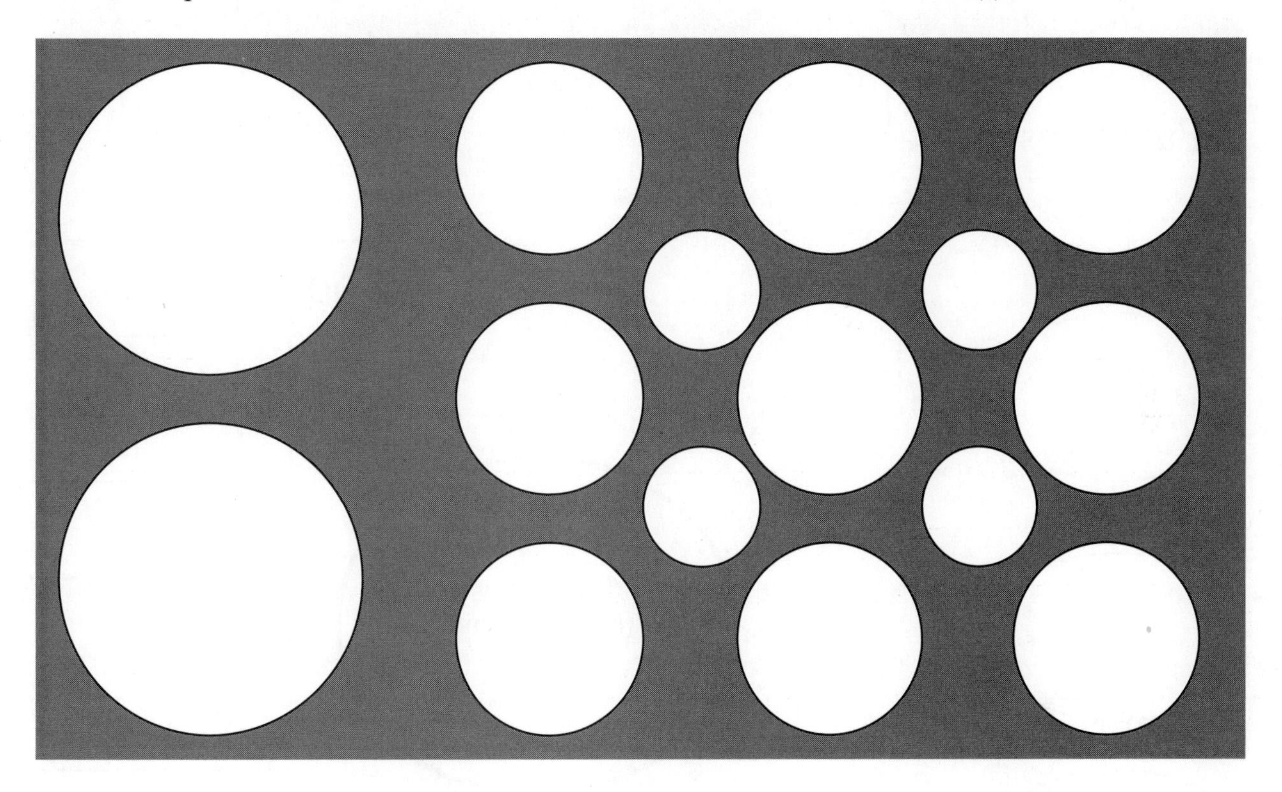

	Ingredients	Minerals
Large Bowl		
Large Bowl		
Medium Bowl		
Medium Bowl		
Medium Bowl		
Medium Bowl		
Medium Bowl		
Medium Bowl		
Medium Bowl		
Medium Bowl		
Medium Bowl		
Small Bowl		
Small Bowl		
Small Bowl		
Small Bowl		

STUDENT WORKSHEET 7-3

TOPIC: Concept Mapping

Mineral Name: ___________________________

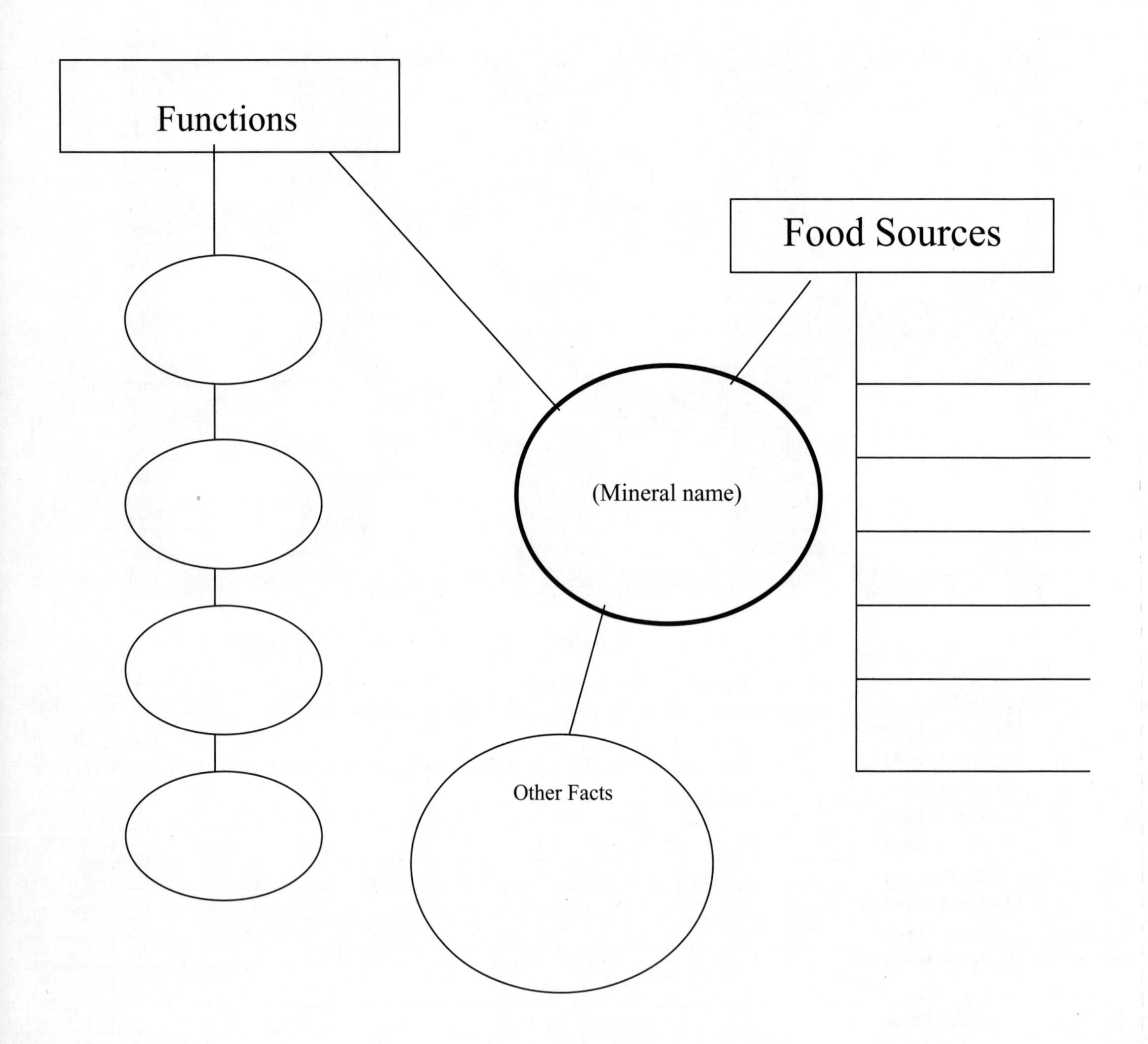

STUDENT WORKSHEET 7-4

TOPIC: Taste Testing of Bottled Water

Have a taste testing of bottled water, including at least two to four products from each of these 3 categories: **spring water**, **purified water**, and **carbonated water**. Compare your results with others in your class.

Brand & Type of Water	Water Quality Notes *	Flavor **	Cost/Cup

* Read about the quality of each water at: **www.bottledwaterweb.com/bott/index.html**

** Taste the water and rate its flavor on a scale of 1 to 5 (1 being poor, 5 being excellent).
Spring water and purified water should taste clean and fresh with no off-tastes or plastic taste.
Carbonated water should be bubbly and a little sour (due to the gas making the bubbles). There will be a touch of bitterness and slight saltiness.

CHAPTER 8 FOUNDATIONS OF HEALTHY COOKING

LEARNING OBJECTIVES

Upon completion of the chapter, the student should be able to:

1. Define seasoning, flavoring, herbs, and spices.
2. Suggest ingredients and methods to develop flavor.
3. Identify and suggest healthy cooking methods and techniques.

CHAPTER OUTLINE

1. **How to develop flavor**

   Seasonings bring out flavor already present in a dish. Flavorings add a new flavor or modify the original one.

   Herbs are leafy parts of certain plants that grow in temperate climates. Spices are roots, bark, seeds, flower, buds, and fruits of certain tropical plants. Herbs and spices are key flavoring ingredients in nutritional menu planning. Examples include:

   - Pepper (black, white, green)
   - Pink peppercorns
   - Red pepper (cayenne)
   - Basil
   - Oregano
   - Tarragon
   - Rosemary
   - Dill, mustard
   - Paprika
   - Chili powder
   - Curry powder
   - Cinnamon, nutmeg, mace
   - Mint

   In addition, you can make seed blends, ethnic blends, as well as toast certain herbs and spices, such as all spice, cumin, fennel, caraway, and coriander.

   Juices can be used as is for added flavor or can be reduced to get more intense flavor, vibrant color, and syrupy texture. Reduced juices make excellent sauces and flavorings.

   Vinegar (wine, cider, balsamic) and oils can add flavor to a wide variety of dishes. Both vinegar and oil can be infused with ingredients for new flavors.

   Stocks—made by simmering water, bones, regular mirepoix, herbs, spices, and sometimes tomatoes. Types of stocks: chicken, fish, crustacean, brown, vegetables. Describe how to make stock. One cup of stock is only about 40 kcalories at most. Thicken stock with arrowroot or cornstarch, or puréed vegetables or potatoes (or simply reduce).

   Glazes are stocks reduced to a thick, gelatinous consistency with flavoring and seasonings. Small amounts of glazes are used to enhance sauces, soups, and other items.

Convenience bases have a high salt content and contain other seasonings and preservatives. This gives them strong and definite tastes that are difficult to work with in building subtle flavors for soups and sauces. Look for bases with the highest-quality ingredients.

Rubs (wet or dry) and marinades also add flavor. To give marinated foods flavor, try citrus zest, diced veggies, fresh herbs, shallots, garlic, low-sodium soy sauce, mustard, and toasted spices.

Aromatic vegetables also add flavor. They include onions, garlic, scallions, leeks, shallots, and chives.

Healthier sauces include vegetable purees, coulis (a sauce made of a puree of vegetables or fruits), salsa (chunky mixtures of vegetables and/or fruits and flavor ingredients), relish (pickles vegetables that are often spicy), chutney (sauce from India that is made with fruits, vegetables, and herbs), compote (a dish of fruit, fresh or dried, cooked in syrup flavored with spices or liqueur), and mojos (a spicy sauce of garlic, citrus juice, oil, and fresh herbs).

Alcoholic beverages and extracts and oils from aromatic plants (vanilla) also add flavor.

Putting it all together: Flavor Profiles. You don't need to separate your cooking methods from healthy to fattening. You should be using the same methods of preparation as you do for any style of cooking. The exceptions are deep-frying, pan-frying, and sautéing with oil. When you are preparing dishes that limit fat, it is smarter to use those fats at the end of your preparation rather than during the cooking process. As you plan each menu item, identify what direction of flavor combinations you want to achieve, such as hints of ginger with soy or garlic with lemon and basil. Your creativity is up to you when you are designing a dish. Mise en place is also very important to create a successful new dish.

2. **Discuss healthy cooking methods and techniques**

Healthy cooking techniques:

- Reduction – boiling or simmering a liquid down to a smaller volume
- Searing – exposing meat's surfaces to a high heat before cooking at a lower temperature, this process adds color and flavor to the meat
- Deglazing – adding liquid to the hot pan used in making sauces and meat dishes, any browned bits of food sticking to the pan are scraped up and added to the liquid
- Sweating – cooking slowly in a small amount of fat over low or moderate heat without browning
- Puréeing – mashing or straining a food to a smooth pulp

Dry heat cooking methods are acceptable when heat is transferred with little or no fat, and excess fat is allowed to drip away from the food being cooked.

- *Roasting*—use a rack so fatty drippings fall to the bottom of the pan; use rubs, marinades, and smoking to develop flavor
- *Broiling and grilling*—both are great ways to bring out flavor from meat, poultry, fish, and vegetables

- *Sauté and dry sauté*—use a well-seasoned nonstick pan and add about half a teaspoon or two sprays of oil (about1 gram of fat) per serving when sautéing, wipe out excess oil for dry sautéing
- *Stir-frying*—mise en place is very important

Moist heat cooking methods do not add the flavor that dry heat methods do. Also fat does not drip away but instead stays in the cooking liquid. To use moist-heat cooking, you need fresh ingredients, seasoned cooking liquids, and strongly flavored sauces or accompaniments.

- *Steaming*—excellent for retaining nutrients in vegetables; fish en papillote retains moisture, flavor, and nutrients
- *Poaching*—court bouillon
- *Braising or stewing*

NUTRITION WEB EXPLORER

***Cooking Light* magazine:** www.cookinglight.com
On the home page of this popular magazine on healthy cooking and living, click on "Cooking 101." Then click on "Techniques," read one of the articles, and write a paragraph about it. Finally, click on "Meet the Chef" and read one of the interviews. Write a summary on the Chef you chose.

The Culinary Institute of America's Professional Chef site: www.ciaprofchef.com
On this home page, click on "World of Flavors," then click on "World Culinary Art Series." Watch the video online for one of the following cuisines: Indian, Spanish, Mexican, Thai, Singapore, Istanbul, or Southern Spain. What type of flavor profile does the cuisine have? Describe below.

Also, click on "Strategies for Chefs" and read one of the articles, such as "Design Menus Seasonally." Summarize the article below.

Flavor-Online.com: www.flavor-online.com

Enter "Flavor Pyramid" in the "Search Articles" box. Read "Building on the Flavor Pyramid." What are the components of the author's flavor pyramid? Describe each one.

***Eating Well* Magazine:** www.eatingwell.com

Click on "Health" and then "Healthy Cooking." Read the article "Tools for a Healthy Kitchen" and list five tools mentioned.

GigaChef.com www.GigaChef.com

Register at this website (it's free), then find five GigaChef recipes that use a variety of flavor builders discussed in this chapter. Write down the titles of the recipes and the flavor builders below.

CHAPTER REVIEW QUIZ

Key Terms: Matching

1. Seasonings
2. Flavorings
3. Herbs
4. Spices
5. Rubs
6. Marinades
7. Coulis
8. Salsas
9. Chutney
10. Compote
11. Mojo
12. Reduction
13. Searing
14. Deglazing
15. Sweating
16. Puréeing

a. A dish of fruit, fresh or dried, cooked in syrup flavored with spices or liqueur; it is often served as an accompaniment or dessert
b. Mashing or straining food to a smooth pulp
c. A sauce made of a purée of vegetables or fruits
d. Adding liquid to the hot pan used in making sauces and meat dishes; any browned bits of food sticking to the pan are scraped up and added to the liquid
e. Substances used in cooking to bring out a flavor that is already present
f. The roots, bark, seeds, flowers, buds, and fruits of certain tropical plants
g. Substances used in cooking to add a new flavor or modify the original flavor
h. A spicy Caribbean sauce; it is a mixture of garlic, citrus juice, oil, and fresh herbs
i. The leafy parts of certain plants that grow in temperate climates; they are used to season and flavor foods
j. Exposing meat's surfaces to a high heat before cooking at a lower temperature; this process adds color and flavor to the meat
k. A dry marinade made of herbs and spices, sometimes moistened with a little oil, that is rubbed or patted on the surface of meat, poultry, or fish
l. Boiling or simmering a liquid down to a smaller volume
m. Chunky mixtures of vegetables and/or fruits and flavor ingredients
n. Cooking slowly in a small amount of fat over low or moderate heat without browning
o. A seasoned liquid used before cooking to flavor and moisten foods; usually based on an acidic ingredient
p. A sauce from India that is made with fruits, vegetables, and herbs

Multiple Choice

1. ____________ comes in three forms: black, white, and green.
    a. Cumin
    b. Garlic
    c. Pepper
    d. Coriander

2. This herb has long, dark green, brittle leaves from a small tree in Asia.
    a. Bay leaves
    b. Anise
    c. Juniper
    d. Fennel seed

3. This is a blend of up to 20 spices that often includes black pepper, cumin, ginger, and cloves.
    a. Jerk
    b. Cayenne
    c. Allspice
    d. Curry

4. Which of the following is the sign of a high-quality stock?
    a. The stock is fat-free
    b. The stock is pleasant to smell and taste
    c. The stock is translucent and free of solid matter
    d. Both b and c
    e. All of the above

5. All of the following vegetables are aromatic except:
    a. Corn
    b. Shallots
    c. Chives
    d. Leeks

6. This is made from fruits, vegetables, and herbs and comes originally from India.
    a. Salsas
    b. Compote
    c. Mojo
    d. Chutney

7. Which of the following is not considered a Mediterranean spice?
    a. Thyme
    b. Oregano
    c. Mustard
    d. Coriander

8. Which of the following is not a whole spice that can be toasted?
    a. Chervil
    b. Cumin
    c. Allspice
    d. Anise

True/False

1. Tarragon has a flavor that tastes something like licorice, and is used in poultry and fish dishes.
    a. True   b. False

2. Sage is a member of the mint family.
    a. True   b. False

3. Rubs can be applied both dry and wet.
    a. True   b. False

4. Moist heating, which is cooking food quickly in a small amount of fat over high heat, can be used to cook tender foods in small portions.
    a. True   b. False

5. Braising involves two steps: searing or browning food in its own fat and then adding liquid and simmering until done.
    a. True   b. False

6. Spices are the leafy parts of certain plants that grow in temperate climates.
    a. True   b. False

7. Dried herbs are preferable over fresh herbs when because they can withstand more than 30 minutes of cooking.
    a. True   b. False

8. Herbs and spices are important for cooking because of their aroma, which contributes approximately 60 percent of their flavor.
    a. True   b. False

Short Answer

1. Which herb has the leaf of an evergreen shrub of the mint family, and looks and feels like pine needles?

Rosemary

2. This herb is a fine powder from a variety of red peppers, and includes both a Hungarian and Spanish variety.

Paprika

3. This orange-yellow root, which is also a member of the ginger family, has a musky, peppery flavor and colors foods yellow.

Tumeric

4. Name five cooking methods and techniques that are used for healthy eating styles.

reduction

searing

deglazing

sweating

purééing

5. How are heterocyclic amines (HCAs) produced?

From amino acids in protein when food are cook at high temps

6. What are polycyclic aromatic hydrocarbons (PAH)?

Cancer-causing substances produced from grilled food falling on hot coals or lava/cerame bricks carried to food by smoke

7. What is a flavored liquid used to make soups, sauces, stews, and braised foods?

Stocks

8. What is the difference between seasonings and flavorings?

s - bring out flavor already present

f - add or modify original flavor

STUDENT WORKSHEET 8-1

Topic: Herbs and Spices

As you examine and learn about specific herbs and spices, write down the following information for each one.

Herb/Spice	Market Forms	Description	Uses

STUDENT WORKSHEET 8-2

Topic: Healthy Cooking Applications

Review this recipe, then list below techniques from chapter 8 that you could use to make this recipe healthier and still maintain flavor.

Lasagna

- 1 pound ground beef
- 1 (14.5 ounce) can Italian stewed tomatoes, cut up
- 1 (6 ounce) can tomato paste
- 1 tablespoon minced fresh parsley
- 1/2 teaspoon minced garlic
- 2 eggs
- 1 1/2 cups small curd cottage cheese
- 1 1/2 cups ricotta cheese
- 1 cup grated Parmesan cheese
- 1 teaspoon salt
- 1 teaspoon pepper
- 6 lasagna noodles, cooked and drained
- 2 cups shredded mozzarella cheese

1. In a large skillet, cook beef over medium heat until no longer pink; drain. Stir in the tomatoes, tomato paste, parsley, and garlic; remove from the heat. In a large bowl, combine the eggs, cheeses, salt, and pepper. Layer three noodles in a greased 13-in. x 9-in. x 2-in. baking dish. Top with half of the cottage cheese mixture, 1 cup mozzarella cheese and half of the meat sauce. Repeat layers.
2. Cover and bake 375 degrees F for 30 minutes. Uncover; bake 25–30 minutes longer or until edges are bubbly. Let stand for 10 minutes before cutting.

<u>Ways to make this lasagna recipe healthier and still retain flavor:</u>

1.

2.

3.

4.

5.

CHAPTER 9 HEALTHY MENUS AND RECIPES

LEARNING OBJECTIVES

Upon completion of the chapter, the student should be able to:

1. Provide your clients with healthy selections in each section of the menu
2. List the elements to consider when presenting foods at their best
3. Select and prepare appropriate garnishes

CHAPTER OUTLINE

1. **Introduction to healthy menus**

   The healthy menu provides nutritious options for guests who want them.

   A healthy meal may be defined as one that includes whole grains, fruits, vegetables, lean protein, and small amounts of healthy oils. Another way to look at it is how many kcalories and nutrients it contains.

   - 800 kcalories or less
   - 35% or fewer kcalories from fat, emphasizing oils high in monounsaturated and polyunsaturated fats
     - 10% or less of total kcalories from saturated fat
     - No trans fat
     - 100 milligrams or less of cholesterol
   - 45 to 65% kcalories from carbohydrates
     - 10 grams or more of fiber
     - 10% or fewer kcalories from added sugars
   - 15 to 25% kcalories from protein
   - 800 milligrams or less of sodium (about 1/3 teaspoon of salt)

   To develop some nutritious menu items, the first step is to look seriously at your menu. Your menu planning may go in one of 3 directions:

   1. Use existing items on your menu.
   2. Modify existing items to make them more nutritious.
   3. Create new selections.

   Keep in mind the following considerations:

   1. Is the menu item tasty? Taste is the key to customer acceptance and the successful marketing of these items. If the food does not taste delicious and have a creative presentation, then no matter how nutritious it may be, it is not going to sell.
   2. Can each menu item be prepared properly by the cooking staff?
   3. Does the menu item blend with and complement the rest of the menu?
   4. Does the menu item meet the food habits and preferences of the guests?
   5. Is the food cost appropriate for the price that can be charged?
   6. Does each menu item require a reasonable amount of preparation time?
   7. Is there a balance of color in the foods themselves and in the garnishes?
   8. Is there a balance of textures, such as coarse, smooth, solid, and soft?
   9. Is there a balance of shape, with different-sized pieces and shapes of food?
   10. Are flavors varied?

11. *Are the food combinations acceptable?*
12. *Are cooking methods varied?*

2. **How to modify recipes**

You can modify a recipe to get less, or more, of kcalories or nutrients.

There are four basic ways to go about modifying a recipe.

1. *Change/add healthy preparation techniques*
2. *Change/add healthy cooking techniques*
3. *Change an ingredient by reducing it, eliminating it, or replacing it*
4. *Add a new ingredient, particularly to add flavor*

Steps to modify a recipe

1. *Examine the nutritional analysis of the product and decide how and how much you want to change the product's nutrient profile. For example, in a meat loaf recipe, you may decide to decrease its fat content to less than 40 percent and increase its complex carbohydrate content to 10 grams per serving.*
2. *Next, you need to consider flavor. What can you do to the recipe to keep maximum flavor? Should you try to mimic the taste of its original version, or will you have to introduce new flavors? Will you be able to produce a tasty dish?*
3. *Next, modify the recipe using any of the methods just discussed. When modifying ingredients, think about what functions each ingredient performs in the recipe. Is it there for appearance, flavor, texture? What will happen if less of an ingredient is used, or a new ingredient is substituted? You also have to consider adding flavoring ingredients in many cases.*
4. *Evaluate your product to see whether it is acceptable. This step often leads to further modification and testing. Be prepared to test the recipe a number of times, and also be prepared for the fact that some modified recipes will never be acceptable.*

Review the examples of how to modify recipes including appetizers, entrees, sauces, dressings, and desserts.

3. **Review chef's tips for each menu section** (pages 320–330)

Breakfast

Appetizers

Soups

Salads and dressings (creamy, puréed, reduction)

Entrees

Side Dishes

Desserts

Morning Breaks

Afternoon Breaks

4. Principles of presentation

Basic principles of presentation:

- Height gives a plate interest and importance.
- Color is very important—but don't overdo it.
- Vary shapes.
- Match the layout of the menu item with the shape of the plate.
- Think about combinations of foods.
- The most effective garnish is bright, eye-catching, contrasting in color, pleasing in shape, and simple in design.

How to make less look like more—fanning, serve larger portions of side dishes, use sauces such as vegetable coulis to help cover the plate and provide eye appeal.

There is a simple method you can use for making attractive garnishes for your menu items. Slice a fruit or vegetable about 1/16th inch thick. Place on silk mats close together without touching. Paint the slices with a thin coat of simple syrup made with water and honey that has been reduced to a light syrup consistency. Place a silk mat over the top and double sheet pan the product. Dry in a 275 degree oven for about an hour or until dry. Fruit and vegetable chips are attractive garnishes. Potatoes work well using this technique, as well as celery root, plantains, pears, apples, and pineapple.

NUTRITION WEB EXPLORER

Carrabba's Restaurants: www.carrabbas.com

Cheesecake Factory: www.thecheesecakefactory.com

Cameron Mitchell Restaurants: www.cameronmitchell.com

Go to the website of any of these restaurants and printout its menu items. Read the descriptions. Circle the menu items that appear to be directed to nutritionally conscious customers. For two of the recipes given in this chapter, write menu descriptions that make the foods sound appealing and let guests know that they are balanced.

The Culinary Institute of America's Professional Chef Site: www.ciaprochef.com
Click on "World of Flavors." Next, click on "Worlds of Healthy Flavors Online," then click on "Profiles, Interviews, and Best Practices." Read how one of the Volume Operators, such as Chartwells or Legal Seafoods, are working healthy options into their operations. Also, click on "Strategies for Chefs" and learn more on how chefs offer healthy dining. Write a summary below.

***Eating Well* Magazine:** www.eatingwell.com
Click on "Recipes," then "Recipe Makeovers" and see how a recipe was made healthier.

CHAPTER REVIEW QUIZ

Multiple Choice

1. A healthy meal contains all the following characteristics except:
    a. 800 or fewer kcalories
    b. 8 grams or more of fiber
    c. 10 percent or fewer kcalories from added sugars
    d. 15 to 25 percent kcalories from protein

2. Which of the following will be altered from an omelet with eggs to an omelet with egg whites?
    a. Cholesterol
    b. Fat
    c. Saturated fat
    d. Both b and c
    e. All of the above

3. Which of the following isn't considered a starchy food?
    a. Legumes
    b. Squash
    c. Yogurt
    d. Peas

4. Simple methods for making attractive garnishes include:
    a. Slicing potatoes
    b. Fruit and vegetable chips
    c. Lattice-style garnishes
    d. Both a and b
    e. All of the above

True/False

1. Pan-searing is another term for dry-sautéing.
    a. True   b. False

2. Half a cup of vegetable oil is a proper and healthy substitution for 1 cup shortening.
    a. True   b. False

3. Mojo is a dish of fruit, fresh or dried, cooked in syrup flavored with spices or liqueur.
    a. True   b. False

4. Color and shape are the two basic principles in food presentation.
    a. True   b. False

5. Bulgur wheat can be used to extend ground meat.
    a. True   b. False

Short Answer

1. Name three diseases that can be prevented, reduced, or reversed with a healthy diet.

   hypertension, obesity + type 2 diabetes

2. What is a healthy recipe substitution for mayonnaise?

   Nonfat yogurt

3. Name three ingredients that add flavor without large amounts of kcalories, fats, or carbohydrates.

   dijon mustard, shallots or garlic, lemon or lime juice

4. Name three different types of salad dressings.

   Creamy, pureed, or reduction dressings

5. Why should reduced-fat cheeses be cooked at lower temperatures for short periods of time?

   too much heat or direct heat may toughen cheeses

6. Why do companies consider afternoon breaks to be important?

   to provide satisfying and well balanced snacks so participants can keep focused on what they are doing (meetings)

STUDENT WORKSHEET 9-1

TOPIC: Recipe Modification

Using any of the techniques described in this chapter, list at least three ways you could modify each recipe below to:

- increase fiber, monounsaturated fats, polyunsaturated fats, and/or
- decrease total fat, saturated fat, cholesterol, sugar, sodium.

If possible, prepare the standard and modified versions to determine if the products are equivalent in taste.

Cheesy Beef Casserole	**Revised Cheesy Beef Casserole**
8 ounces elbow macaroni 2 tablespoons fat 1 pound ground chuck ½ small onion, finely chopped 10 ounce can cheddar cheese soup 10 fluid ounces milk ½ teaspoon salt ¼ teaspoon pepper 2 slices breads made into crumbs 1. Boil elbows for 8 minutes. Drain. 2. Brown beef and onion, drain off fat. 3. Add soup and then milk slowly to meat mixture. Add salt and pepper. Stir in macaroni. 4. Put meat mixture into 2 quart casserole. Cover with bread crumbs 5. Bake at 350°F for 15 minutes until bubbly.	
Carrot-Raisin Salad	**Revised Carrot-Raisin Salad**
2 cups shredded raw carrots ¼ cup seedless raisins ½ cup mayonnaise 2 tablespoons fresh lemon juice dash salt Combine all ingredients and chill.	

Chocolate Chip Cookies	Revised Chocolate Chip Cookies
2¼ cup all-purpose flour ½ cup quick-cooking oats 1 teaspoon baking soda ½ teaspoon salt 1 cup butter 1 cup firmly packed brown sugar 2 teaspoons vanilla extract 2 cup (12 oz.) semi-sweet chips 2 large eggs ¾ cup granulated sugar 2 cups chopped nuts 1. Preheat oven to 375°F. 2. Combine flour, oats, baking soda, and salt in bowl. 3. Beat butter and sugar in large mixing bowl at medium speed until creamy. 4. Add eggs one at a time, mixing well after each addition. Beat in vanilla. 5. At low speed, beat in flour mixture until well blended. 6. With a wooden spoon, stir in chocolate chips and nuts. 7. Drop 1 heaping tablespoon of dough 2 inches apart on ungreased cookie sheets. 8. Bake 13 to 14 minutes until golden brown. 9. Let cookies rest 2 minutes then transfer to wire rack to cool.	
French Dressing	**Revised French Dressing**
½ cup oil 2 tablespoons vinegar ½ teaspoon salt ½ teaspoon paprika ½ teaspoon sugar Combine all ingredients and mix well.	

CHAPTER 10 MARKETING TO HEALTH-CONSCIOUS GUESTS

LEARNING OBJECTIVES

Upon completion of the chapter, the student should be able to:

1. Describe two methods a foodservice operator can use to gauge customers' needs and wants
2. Give three examples of ways to draw attention to healthy menu options
3. Discuss effective ways to communicate and promote healthy menu options
4. Explain the importance and extent of staff training needed to implement healthy menu options
5. Describe two ways to evaluate healthy menu options
6. Respond with menu ideas for special requests from guests
7. Discuss how nutrition labeling laws regulate nutrient content or health claims on restaurant menus

CHAPTER OUTLINE

1. **Marketing**
   Define marketing (the process of finding out what your customers need and want, and then developing, promoting, and selling the products and services they desire.

2. **Gauging customers' needs and wants**
   Most operators who have successfully implemented healthy menu options have done so through reviewing eating trends, examining what other operators are doing, and keeping abreast of their customers' requests for healthy foods.

   Methods
   - *Waitstaff feedback*
   - *Customer surveys*

3. **Adding healthy menu options to the menu**
   People involved: operators, directors, managers, chefs, cooking staff, nutrition experts.

   There are different ways to communicate the menu items.
   - Simply give a good description of your menu item
   - Use the waitstaff to offer and describe nutritious menu options (Fig 10-4 and 10-5 show menus from popular quality restaurant chains that have a wide variety of menu items that can be mixed and matched or served with substitutes to accommodate special requests from diners.)
   - Highlight nutritious menu selections with symbols or words such as "light"
   - Include a special, separate section on the regular menu
   - Add a clip-on to the regular menu and/or a blackboard/lightboard.

   Keep in mind that customers generally don't want kcalorie counts, fat grams, or other such information on the menu. Give a good description of the ingredients, portion size, and preparation method. Market healthy menu items in a positive manner.

4. **Promotion of healthy options**
   There are three methods for promoting healthy options.
   - *Advertising*
   - *Sales promotion—such as coupons, point-of-purchase displays, contests*
   - *Publicity—such as sending press releases, writing a column for a local paper, offering cooking lessons/demonstrations, and so on*

5. **Train the staff**
   Staff training centers on waitstaff and cooking staff

   The waitstaff needs to understand the following:
   - *The scope and rationale for the nutrition program*
   - *Grand opening details*
   - *The ingredients, preparation, and service for each menu item*
   - *Some basic food and nutrition concepts so they can help guests with special dietary needs such as food allergies*
   - *How to handle special customer requests such as the availability of half portions*
   - *Merchandising and promotional details*

   The cooking staff needs training on:
   - *The scope and rationale for the nutrition program*
   - *Grand opening details*
   - *The ingredients, preparation, portion size, and plating of each new menu item*
   - *Some basic food and nutrition concepts so they can help guests with special dietary needs such as food allergies*
   - *How to respond to special dietary requests*

   Training the cooking staff to prepare healthy dishes correctly can be challenging.

   Close supervision is necessary during opening night.

6. **How to evaluate your program**
   1. How did the program do operationally? Did the cooks prepare and plate foods correctly? Did the waitstaff promote the program and answer questions well?

   2. Did the food look good and taste good?

   3. How well did each of the healthy menu options sell? How much did each item contribute to profits? How did the overall program affect profitability?

   4. Did the program increase customer satisfaction? What was the overall feedback from customers? Did the program create repeat customers?

7. **How to respond to special requests**
   To respond to special guest requests, keep in mind these basic preparations:

   - When marinating meats there are many no-salt, no-sugar rubs and seasonings that can be used. The addition of salt to your proteins can be done at time of cooking (or not at all), so a request for no salt is easy to accommodate.
   - Blanched vegetables should be reheated in a small amount of seasoned stock, then finished with whole butter, an extra virgin olive oil, or flavored nut oil. These delicate oils will be the first flavor your customers will taste, giving a rich body and the taste fulfillment up front with fewer fats then before.
   - Dips and chips are and always will be an American favorite for appetizers. This can still be a great selection to your menu with a new twist. Hummus, baba ghanoush, white bean and roasted garlic, artichoke, and goat cheese are well-accepted favorites dips that can be accompanied by baked whole-wheat tortilla chips, melba toast, baked multigrain croutons, or a variety of different vegetables.
   - Create a well-balanced dressing that is low in fat and made with extra virgin olive oil and good vinegars as we have suggested, and finished with fresh herbs and spices. Use this as one of your house dressings so it is available to prepare a number of different choices.
   - Keep a stock or clear broth for reheating vegetables, because many of the new eating habits today recommend starting your meal with a cup of clear broth. This helps to curb the appetite. This can be a kitchen staple as well as quick-serve dual purpose inexpensive starter on your menu.
   - Desserts, desserts, desserts: Creating balanced desserts that your guests can enjoy should not cause stress. There are more than enough ways to add limited sugar to a menu item that can appeal to the majority of eating styles today. A ricotta cheesecake with a roasted walnut, spices, and Splenda crust; a flourless chocolate cake with fresh fruit garnish; a toasted oatmeal, chocolate, banana, pecan pudding with a kiwi or mixed fruit salsa; a buttery phyllo cylinder with maple cream pineapple chutney and berries in season; a banana polenta soufflé[as] with chocolate sauce and glazed banana slices; or an old fashion berry shortcake with fruit sauce berries and whipped cream.

   Use Figure 10-10 on pages 375–377 to find appropriate menus choices for these diets.
   Diet low in fat, saturated fat, and cholesterol
   Low-sodium diet
   Vegetarian diet
   High fiber diet
   Low-lactose diet
   Gluten-free diet
   Diet low in added sugars

8. **Vegetarian eating styles**
   Vegetarians do not eat food that requires the death of, or injury to, an animal.
   Largest group is lacto-ovovegetarians (eat eggs, milk and milk products).

   Lactovegetarians consume milk and milk products but forego eggs.

Vegans do not eat eggs or dairy products—only 4% of vegetarians are vegans.

Pescovegetarians eat seafood.

9. **Why individuals become vegetarians**
   Health reasons: Vegetarians tend to be at lower risk for:
   - hypertension
   - coronary artery disease
   - colon and lung cancer
   - Type 2 diabetes mellitus
   - diverticular disease of the colon

   Ecology: Livestock and poultry require much land, energy, water, and plant food to be produced—this is wasteful

   Economics: A vegetarian diet is less expensive

   Ethics: Animals should not suffer or be killed unnecessarily

   Religious beliefs

10. **Discuss the nutritional adequacy of the vegetarian diet**
    Most vegetarians get enough protein and their diets are typically lower in fat, saturated fat, and cholesterol.

    Some nutrients that need special attention.

    Vitamin B1: Because it is found only in animal foods, vegans need wither a supplement or B12-fortified foods, such as most ready-to-eat cereals, meat analogs, and so on.

    Vitamin D: Because it is found mostly in milk, vegans without enough sunlight exposure may need a supplementary source of vitamin D.

    Calcium: Vegans may have a problem here if they don't eat enough calcium-rich foods such as calcium-fortified orange juice or tofu made with calcium.

    Iron: Vegetarians don't have any more problems with iron-deficiency anemia than meat-eaters, but vegetarians need to keep in mind that vitamin C foods enhances absorption of iron from plant sources.

    Zinc: Zinc is found in many plant foods.

    Infants, children, and adolescents can follow vegetarian diets, and even vegan diets. For growing youngsters, however, vegetarian diets need to be well planned, varied, and adequate in kcalories. In the case of a vegan diet, the focus is on getting enough kcalories, vitamins D and B12, calcium, iron, zinc, and linolenic acid.

11. **Discuss the Vegetarian Food Pyramid and menu planning guidelines**
    Discuss the vegetarian food pyramid (Fig 10-12).

    Menu planning guidelines
    1. *Use a variety of plant protein sources at each meal.*
    2. *Use a wide variety of vegetables.*
    3. *Choose low-fat and nonfat varieties of milk and milk products and limit the use of eggs.*
    4. *Offer dishes made with soybean-based products such as tofu and tempeh.*
    5. *For menu ideas, don't forget to look at cuisines from other countries (Fig 10-13).*

12. **Restaurants and nutrition labeling laws**
    Food prepared and served in restaurants or other foodservices are exempt from mandatory nutrition labeling found in packaged foods.

    Restaurants are not exempt from FDA rules concerning nutrient claims and health claims (discussed in Chapter 2) when used on menus, table tents, posters, or signs. Any food being used in a health claim may not contain more than 20% of the Daily Value for fat, saturated fat, cholesterol, or sodium.

    When providing nutrition information for a nutrient or health claim, restaurants do not have to provide the standard nutrition information profile and more exacting nutrient content values required in the Nutrition Facts panel of packaged foods. They can present the information in any format desired, and they have to provide only information about the nutrient(s) that the claim is referring to.

    Restaurants may use symbols on the menu to highlight the nutritional content of specific items. They are required to explain the criteria used for the symbols.

NUTRITION WEB EXPLORER

Food marketing boards and associations:

American Egg Board: www.aeb.org

Mann Packing: www.broccoli.com

Grains Nutrition Information Center: www.wheatfoods.org

Milk: www.whymilk.com

National Pork Producers Council: www.nppc.org

National Turkey Federation: www.eatturkey.com

Produce Marketing Association: www.pma.com

Pick one of these food marketing board/association websites to visit. Write a brief report about the website, including the name of the board/association, the website address, and a list of items available that a foodservice operator could use (such as recipes), and attach sample material.

Restaurants: www.rockfishseafood.com

www.marcellasristorante.com

www.pfchangs.com

Look at the menus at these restaurant websites and find menu items that you could use to meet guest special requests. Which restaurant chain provides the easiest-to-use nutritional information?

CHAPTER REVIEW QUIZ

Key Terms: Matching

1. Marketing
2. Advertising
3. Sales promotion
4. Publicity
5. Press release

a. A printed announcement by a company about its activities, written in the form of a news article and given to the media to generate publicity
b. Any paid form of calling public attention to the goods, services, or ideas of a company or sponsor
c. The process of finding out what your customers need and want and then developing, promoting, and selling the products and services they desire
d. Obtaining free space or time in various media to get public notice of a program, book, and so on
e. Marketing activities other than advertising and public relations that offer an extra incentive

Multiple Choice

1. _____________ involves obtaining free editorial space or time in various media.
    a. Sales promotions
    b. Publicity
    c. Advertising
    d. Marketing

2. When catering to health-conscious guests, the cooking staff's training must include:
    a. How to respond to special dietary requests
    b. Grand opening details
    c. Ingredients, preparation, portion size, and plating of each menu item
    d. The scope and rationale for the nutrition program
    e. All of the above

3. How many kcalories are in 1 tablespoon of oil?
    a. 60
    b. 90
    c. 100
    d. 120

4. Important nutrients for vegetarians are:
    a. Vitamin D
    b. Vitamin B12
    c. Zinc
    d. Iron
    e. All of the above

5. Which of the following are unique in that they contain plant protein that is nutritionally equivalent to animal protein?
   a. Whole wheat
   b. Soybeans
   c. Kidney beans
   d. Peanuts

6. A group of vegetarians who do not eat eggs or dairy products are called:
   a. Lacto ovo vegetarians
   b. Lacto vegetarians
   c. Pesco vegetarians
   d. Vegans

7. Lactose is present in large amounts in all but:
   a. Yogurt
   b. Eggnog
   c. Cottage cheese
   d. Ice milk

8. Which of the following won't trigger a problem for a person with celiac disease?
   a. Rice
   b. Barley
   c. Rye
   d. Wheat

9. Processed foods contribute what amount of sodium in most people's diets?
   a. 50%
   b. 25%
   c. 75%
   d. 60%

True/False

1. Marketing is a paid form of calling public attention to the goods, services, or ideas of a company or sponsor.
   a. True b. False

2. All low-carbohydrate diets emphasize protein and non-starchy vegetables.
   a. True b. False

3. Legumes are part of a high fiber diet.
   a. True b. False

4. Monounsaturated fat can help decrease cholesterol levels.
   a. True b. False

5. Celiac disease is an inherited autoimmune disease.
   a. True   b. False

6. To be considered a good source of calcium, a food item must provide more than 20 percent of the Daily Value for calcium in one serving.
   a. True   b. False

7. Restaurants must have nutrition information available upon request for any menu items that contain nutrient or health claims.
   a. True   b. False

8. Salt by weight is 60 percent sodium.
   a. True   b. False

Short Answer

1. Where do the major sources of added sweeteners in the American diet come from?

2. List four staples of a gluten-free diet.

3. Which types of cheese cause the fewest problems for people with lactose intolerance?

STUDENT WORKSHEET 10-1

TOPIC: Healthy Menu Options

Italian restaurants are very popular across the United States. Following are menu items from a traditional Italian restaurant. For each category of the menu, you need to develop one healthy menu option (it may already be on the menu). How would you promote them?

Pizza	*Calzone and Stromboli*
Pan Pizza (medium or large) Thin Crust Pizza (medium or large) Sicilian Pizza (medium or large) Toppings: Sausage, mushrooms, anchovies, pepperoni, onions, garlic, green peppers, black olives, clams, shrimp, bacon, eggplant, sundried tomatoes, extra cheese.	Small Calzone Large Calzone Sausage Roll with Cheese and 2 toppings Toppings: ham, salami, pepperoni, sausage
Appetizers	***Spaghetti, Linguine, or Ziti with***
Homemade Clams Casino Roasted Pepepr Garlic Bread Mussels Marinara Mozzarella Caprese (fresh mozzarella, sliced tomatoes, olives, and olive oil) Brushetta (diced tomatoes, garlic, olive oil, basil and spices) Garlic Bread Garlic Bread with Cheese Fried Shrimp Mozzarella Sticks Jalapeno Poppers Calamari Rings Buffalo Wings (12 or 24)	Meatballs Sausage Meat Sauce White or Red Clam Sauce Mushroom Sauce Marinara Sauce Oil and Garlic Tomato Sauce Peso Sauce Broccoli in Oil and Garlic Sauce Shrimp Cooked in Marinara Sauce Sun-Dried Tomatoes and Gaeta Olives Olive Oil, Garlic Sauce and Capers Clams and Mussels in a Marinara Sauce
Baked Dishes	***Dinners***
Gnocchi Stuffed Shells Manicotti Ravioli, Cheese or Meat Meat Lasagna Vegetable Lasagna Ziti	Veal, Eggplant, Shrimp, or Chicken Parmigiana New York Strip Steak with French Fries or Onion Rings Veal or Chicken Francaise with Pasta and Salad Veal or Chicken Marsala with Pasta and Salad Lobster Ravioli in Pink Cream Sauce and Salad Stuffed Rigatoni with Salad

Salads	***Side Orders***
Tossed Salad Antipasto Tuna Salad Caesar Salad	French Fries Cheese Fries Onion Rings Sausage Meatballs
Hot Subs	***Cold Subs***
Chicken Parmigiana Veal Parmigiana Eggplant Parmigiana Sausage Parmigiana Meatball or Sausage Sausage & Peppers Cheeseburger Italian ¼ Pound Hot Dog	Italian Hoagie Homemade Roast Pork Hoagie Homemade Roast Beef Hoagie Turkey Hoagie Tuna Hoagie Ham and Cheese Hoagie

HEALTHY MENU OPTIONS

Pizza:

Calzone and Stromboli:

Appetizers:

Spaghetti, Linguine, or Ziti:

Baked Dishes:

Dinners:

Salads:

Side Orders:

Hot/Cold Subs:

STUDENT WORKSHEET 10-2

TOPIC: Taste Test of Vegetarian Burgers

The number and popularity of vegetarian burgers has been increasing tremendously. Many restaurants, even Burger King, offer vegetarian burgers on the menu. Use this sheet to write down information about the vegetarian burgers you will taste.

Brand Name	Serving Size	Major Ingredients	Kcal	Grams Fat	Taste*	Comments
1.						
2.						
3.						
4.						
5.						
6.						
7.						
8.						
9.						

- Rank taste on a scale of 1 to 5.

  1 – Excellent
  2 – Very Good
  3 – Good
  4 – Fair
  5 – I would only eat this if I was VERY hungry.

STUDENT WORKSHEET 10-3

TOPIC: Vegetarian Meal

Below you will find recipes for two vegetarian main dishes and two desserts. Most recipes use a version of soy.

Easy Tofu Lasagna
(9 servings)

8 ounces mushrooms, chopped
1½ cups zucchini, chopped
16 ounces tofu
1 tablespoon lemon juice
1 tablespoon dried parsley flakes
1 teaspoon Italian herb seasoning
¼ teaspoon black pepper
¾ cup water
4 cups fat-free marinara sauce
8 ounces lasagna noodles, uncooked
4 ounces Mozzarella-style soy cheese
¼ cup Parmesan-style soy cheese

1. Preheat the oven to 350°F.
2. Sauté the mushrooms and zucchini in a non-stick skillet until tender, adding a little water if needed. Set aside.
3. Mash the tofu in a small mixing bowl. Add the lemon juice and the 3 seasonings. Mix well.
4. Combine the water and marinara sauce in a bowl.
5. Assemble the lasagna: put about 1/3 of the sauce on the bottom of a 9" x 13" baking dish. Top with half the uncooked noodles, half the tofu mixture, half the Mozzarella-style soy cheese, and all of the mushrooms and zucchini. Put another 1/3 of the sauce on top, the remaining noodles, the remaining tofu and then the last 1/3 of the sauce. Top with the remaining cheeses.
6. Cover the casserole with foil. Bake at 350°F for 1 hour.
7. Remove from oven and let sit 10 minutes before serving.

Vanilla Pudding
(3 servings)

½ cup sugar
2 tablespoons cornstarch
1½ cups plain soymilk
1 teaspoon vanilla extract

1. Stir the sugar and cornstarch together in a medium saucepan. Whisk in the soymilk.
2. Cook, stirring, over moderate heat until the mixture comes to a boil and thickens.
3. Remove from heat, stir in vanilla, and pour into 3 individual serving dishes. Chill.

Creamy Lentil & Mushroom Stew
(4–6 servings)

1 quart water
1 cup dry lentils
1 large potato, chopped
4 ounces fresh spinach
2 tomatoes, chopped
2 tablespoons oil
1 garlic clove, minced
1 onion, chopped
½ pound mushrooms, sliced
1 cup dry noodles, cooked al dente
1 teaspoon oregano
½ teaspoon curry
1 cup plain yogurt
1 cup cottage cheese
1 tablespoon honey
Salt and pepper to taste

1. Bring the water to a boil in a large saucepan. Add the lentils. Cover and simmer for about 25 minutes. When the lentils begin to soften, add the potato, spinach, and tomatoes. Keep the mixture covered except to stir. Add a little water if the stew begins to get too thick.
2. Heat the oil in a large skillet. Add the garlic, onion, and mushrooms and sauté until nicely browned. Combine with the lentil mixture and add the remaining ingredients.
3. Simmer the stew until it reaches the desired consistency. Serve.

Michigan Fruit Crisp
(6 servings)

Filling:
2 cups fresh cherries, pitted and halved
2 cups fresh blueberries
2 tablespoons flour
½ cup sugar
½ teaspoon cinnamon

Topping:
1/3 cup dry TSP (textured soy protein)
½ cup oats
¼ cup brown sugar
2 tablespoons flour
¼ teaspoon cinnamon
2 tablespoons margarine, melted

1. Preheat the oven to 350°F.
2. Make the filling: put the fruit in a medium mixing bowl. Combine the flour, sugar, and cinnamon and pour over the fruit. Stir together well. Pour into an 8" or 9" square baking dish.
3. Make the topping: combine all topping ingredients except the margarine in a small mixing bowl. Add the margarine and stir to combine well. Using your fingers, sprinkle the topping over the fruit.
4. Bake for 45 minutes, until the filling bubbles around the edges.

CHAPTER 11 NUTRITION AND HEALTH

LEARNING OBJECTIVES

Upon completion of the chapter, the student should be able to:

1. List and describe three common forms of cardiovascular disease
2. Explain what atherosclerosis is and how it is related to cardiovascular diseases
3. List five risk factors for coronary heart disease
4. Distinguish between angina and a heart attack
5. Explain how a person's risk for coronary heart disease is assessed
6. Explain the two main ways to lower blood cholesterol levels
7. Explain how strokes occur
8. List five lifestyle modifications for hypertension control
9. List five menu-planning guidelines to lower cardiovascular risk
10. Define cancer
11. Outline the American Cancer Society's four guidelines to reduce cancer risk
12. Distinguish between type 1 and type 2 diabetes mellitus and understand the principles of planning meals for people with diabetes
13. Define osteoporosis and how to prevent/treat it
14. Discuss how to safely use botanicals including herbs
15. Analyze the pros and cons of biotechnology used to produce plants for food

CHAPTER OUTLINE

1. **Introduction**

   The 2 leading causes of death in the U.S. are cardiovascular disease and cancer. These diseases have one thing in common: diet.

2. **Nutrition and cardiovascular disease (CVD)**

   Cardiovascular disease is a general term for diseases of the heart and blood vessels, such as:

   - Coronary artery disease
   - Stroke
   - High blood pressure

   The two medical conditions that lead to most CVD are atherosclerosis (plaque buildup in the arteries) and high blood pressure. Both are silent.

   Coronary heart disease is a broad term used to describe damage to or malfunction of the heart caused by narrowing or blockage of the coronary arteries.

   Smoking, high blood cholesterol, and high blood pressure are 3 major risk factors for coronary heart disease. Other risk factors include cigarette smoking, physical inactivity, overweight/obesity, and diabetes. Risk factors you can't control are increasing age, male gender, and a family history.

   People with at least three of the following conditions in what is known as metabolic syndrome are at increased risk of dying from coronary heart disease and cardiovascular diseases.

- excessive abdominal obesity: over 40 inches for men and over 35 inches for women
- high blood triglycerides
- reduced high-density lipoprotein (HDL)
- elevated fasting glucose
- high blood pressure

Angina refers to symptoms of pressing, intense pain in the heart area caused by insufficient blood flow to the heart muscle. Most heart attacks (myocardial infarction) are caused by a clot in a coronary artery at the site of narrowing and hardening that stops the flow of blood. Coronary heart disease is the number-one killer of women—heart attacks occur later in women than men.

Your risk for coronary heart disease (CHD) can be assessed by measuring total blood cholesterol as well as the proportions of the various types of lipoproteins. Low-density lipoprotein (LDL) is known as "bad cholesterol" because it increases CHD risk. High-density lipoprotein (HDL) protects the arteries by taking cholesterol to the liver for disposal.

- Total blood cholesterol level of less than 200 mg/dL is desirable
- LDL of less than 100 mg/dL is optimal
- HDL less than 40 mg/dL is low

The main goal of cholesterol-lowering treatment is to lower LDL enough to reduce risk of developing heart disease. There are two main ways to lower cholesterol:

1. Therapeutic lifestyle changes (TLC): includes a cholesterol-lowering diet (TLC diet), physical activity, and weight management
2. Drug treatment: if needed

TLC diet: low saturated fat (less than 7% of total kcal), low cholesterol (less than 200 mg/day), enough kcal to maintain a desirable weight, soluble fiber and plant stanols or sterols may be used to decrease cholesterol further.

A stroke is damage to brain cells resulting from an interruption of the blood flow to the brain. Most strokes are caused by blockages in the arteries that supply blood to the brain. Stroke and heart disease have some of the same controllable risk factors: high blood pressure, cigarette smoking, high cholesterol, diabetes, physical inactivity, and obesity.

3. **High blood pressure**

Because high blood pressure usually doesn't give early warning signs, it is known as the silent killer.

Hypertension is a major risk factor for heart disease and strokes.

Whether your blood pressure is high, low, or normal depends mainly on several factors: the output from your heart, the resistance to blood flow by your blood vessels, the volume of your blood, and blood distribution to the various organs.

Everyone experiences hourly and even moment-by-moment blood pressure changes. For example, your blood pressure will temporarily rise with strong emotions such as anger and frustration. These transient elevations in blood pressure usually don't indicate disease or abnormality.

Blood pressure is spoken of as a fraction, such as 120/80 millimeters of mercury (mmHg). The numerator (120) is called the systolic pressure—the pressure of blood within arteries when the heart is pumping. The denominator (80) is called the diastolic pressure—the pressure in the arteries when the heart is resting between beats. Systolic blood pressure is the key determinant for assessing the presence and severity of high blood pressure for middle-aged and older adults. Normal blood pressure is less than 120/80.

Most cases of hypertension are called essential or primary hypertension and its cause is unknown.

The prevalence of high blood pressure increases with age.

The following lifestyle modifications offer some hope for prevention of hypertension and are effective in lowering blood pressure:

- Maintain normal body weight, lose weight if overweight
- Adopt the DASH eating plan
- Reduce dietary sodium intake to no more than 2,400 mg
- Engage in regular aerobic physical activity
- Limit consumption of alcohol to no more than 2 drinks per day for men and 1 drink for women and lighter-weight persons

The DASH diet recommends 8 to 10 servings of fruits and vegetables (good sources of potassium, vitamin C, and magnesium) and 2–3 servings of dairy products (calcium) daily. The DASH diet also emphasizes low fat, saturated fat, cholesterol, and added sugars.

If lifestyle modifications don't lower blood pressure enough, drug treatment is the next choice.

4. **Menu planning for cardiovascular diseases**
   Menu planning for cardiovascular diseases revolves around offering dishes rich in complex carbohydrates and fiber and using small amounts of fat, saturated fat, cholesterol, and sodium.

5. **Nutrition and cancer**
   Cancer is the second leading cause of death in the US.

   Cancer is a group of diseases characterized by unrestrained cell division and growth that can spread beyond the tissue in which it started. Cancer is a two-step process. First an initiator (such as x-rays, smoking, certain chemicals) alters the genetic material of a cell and causes a mutation. Often these cells are repaired or replaced, but when this does not occur, promoters (such as alcohol or fat) can advance development of the mutated cell into a tumor.

Cancer develops as a result of interactions between internal (genetics, hormones) and external (tobacco, environmental, etc.) factors. Whereas smoking causes the greatest number of cancer cases, up to one-third of all cancer deaths are possibly related to nutrition, overweight or obesity, and physical inactivity.

ACS guidelines to reduce cancer risk:

1. Maintain a healthful weight throughout life.
   Balance kcaloric intake with physical activity
   Avoid excessive weight gain as you get older
   If you are currently overweight or obese, achieve and maintain a healthy weight.
2. Adopt a physically active lifestyle
   Adults: Engage in at least 30 minutes of moderate to vigorous physical activity, above usual activities, on 5 or more days of the week; 45 to 60 minutes of intentional physical activity are preferable
   Children and adolescents: Engage in at least 60 minutes per day of moderate to vigorous physical activity at least 5 days per week
3. Eat a healthy diet with an emphasis on plant sources
   Eat 5 or more servings of a variety of vegetables and fruits each day
   Choose whole grains in preference to processed (refined) grains
   Limit consumption of processed and red meats
4. If you drink alcoholic beverages, limit consumption to no more than 1 drink per day for women or 2 per day for men

Some fruits and vegetables, as well as other plant foods, contain phytochemicals—minute plant compounds that fight cancer formation. Members of the cabbage family, also called cruciferous vegetables, contain phytochemicals such as indoles. They activate enzymes that destroy carcingogens.

Menu planning to lower cancer risk:
1. Offer lower-fat menu items
2. Avoid salt-cured, smoked, and nitrite-cured foods
3. Offer high-fiber foods
4. Include lots of vegetables, especially cruciferous vegetables
5. Offer foods that are good sources of beta carotene and vitamins C and E
6. Offer alternatives to alcoholic drinks
7. See guidelines for grilling foods on page 287

6. **Caffeine**

Caffeine is a stimulant found in coffee, tea, cola, cola, and medications. It is easy to become dependent on caffeine. When caffeine is withdrawn, symptoms include headache, fatigue, irritability, depression, and poor concentration. Caffeine is not considered a diuretic.

Moderate use of caffeine is considered to be 300 milligrams, which is equal to about 3 cups of coffee daily.

Caffeine appears to increase the excretion of calcium. Moderate caffeine intake doesn't seem to cause a problem with calcium as long as you are consuming the recommended amounts.

Caffeine's ability to improve physical performance is well known among well-trained athletes.

7. **Discuss nutrition and diabetes mellitus**

Diabetes mellitus is a disorder of carbohydrate metabolism characterized by high blood sugar levels and inadequate or ineffective insulin.

The number of people diagnosed with diabetes has almost doubled since 1990.

The life expectancy for a diabetic is only two-thirds that of the non-diabetic. People with diabetes are more than ordinarily vulnerable to many kinds of infections and to deterioration of the kidneys, heart, blood vessels, nerves, and vision.

There are 2 types of diabetes:

- *Type 1—seen mostly in children and adolescents, insulin injections are needed because the person makes none, less than 10% of persons with diabetes have Type I*
- *Type 2—seen in older, usually obese adults, these adults make insulin but their tissues aren't sensitive enough to insulin, treatment is with diet, weight reduction when needed, exercise, and sometimes oral medications that improve the body's sensitivity to insulin*

Symptoms of Type 1 diabetes typically appear abruptly and include excessive, frequent urination, insatiable hunger, unquenchable thirst, and the formation of ketone bodies. Unexplained weight loss is also common, as are blurred vision (or other vision changes). Symptoms of Type 2 are similar, but they come on much more slowly, and the problem with ketone bodies is rare.

Treatment for either type seeks to do what the human body normally does naturally: maintain a proper balance between glucose and insulin. The guiding principle is that food makes the blood glucose level rise while insulin and exercise make it fall. The trick is to juggle the three factors to avoid both hyperglycemia, meaning a blood glucose level that is too high, and hypoglycemia, meaning one that is too low.

The current diabetic diet is based on the following principles:

- *There is no one diet suitable for every person with diabetes*
- *The goals are to maintain the best glucose control possible, keep blood levels of fat and cholesterol in normal ranges, maintain or get body weight within a desirable range, and meet all nutrient needs*
- *Sugar can be incorporated into the diet as part of the total carbohydrate allowance*
- *Instead of setting rigid percentages of protein, fat, and carbohydrates, guidelines recommend that 60 to 70% of kcalories should come from carbohydrates and monounsaturated fats. Saturated fats should be kept at 10% or less. The meal pattern should provide sufficient fiber. Kcaloric distribution depends on the individual's nutritional assessment and treatment goals.*

Explain how the diabetic exchange lists are used by people with diabetes.

Starch—1 slice bread, 80 kcalories
Meat—1 ounce lean meat, 55 kcalories
Vegetable—½ cup cooked vegetables, 25 kcalories
Fruit—1 small apple, 60 kcalories
Milk—1 cup nonfat milk, 90 kcalories
Other carbohydrates—2 small cookies, kcalories vary
Fat—1 teaspoon margarine, 45 kcalories

8. **Osteoporosis**

Osteoporosis is characterized by loss of bone density and strength and debilitating fractures especially in people over 45, due to a tremendous loss of bone tissue in midlife.

Peak bone mass is attained during the early thirties. Bone is being constantly remodeled, and after the early thirties, bone is broken down faster than it is deposited (especially during the five years after menopause for women due to estrogen being greatly decreased).

Bone health is influenced by gender, age, body size, ethnicity, family history (factors you can't change). Risk factors you can change include a diet low in calcium and vitamin D, sedentary lifestyle, cigarette smoking, or excessive use of alcohol.

The best approach to osteoporosis is prevention – taking in the AI for calcium, regular exercise, consuming milk for adequate vitamin D, consuming moderate amounts of alcohol, and avoiding smoking. Medications, such as Fosamax, can increase bone density.

9. **Food Facts: Botanicals and Herbs**

Botanicals are plans or plant parts valued for its medicinal or therapeutic properties, flavor, and/or scent. Herbs are a category of botanicals. Products made from botanicals that are used to maintain or improve health may be called herbal or botanical products. Some botanicals are sold as dietary supplements.

Many people believe that products labeled "natural" are safe. This is not always true because the safety of a botanical depends on many things, such as its chemical makeup, how it works in the body, how it is prepared, and the dose used. In the US, herbal and other botanical supplements are regulated by the US Food and Drug Administration as foods. The actions of botanicals range from mild to powerful.

Herbal supplements can act in the same way as drugs. Therefore they can cause medical problems if not used correctly. The active ingredient(s) in many herbs and other botanical supplements are not known.

10. **Hot Topic: Biotechnology**

Biotechnology is a collection of scientific techniques, including genetic engineering, that are used to create, improve, or modify plants, animals, and microorganisms. **Genetic engineering** is a process in which genes are transferred to a plant, animal, or microbe to have a certain effect, such as produce a soybean with built-in insecticide. Farmers and scientists have been genetically modifying plants for hundreds of years, most often using a process called crossbreeding.

Each cell contains a complete copy of your genetic plan or blueprint. The language of DNA is common to all organisms, whether microbes, plants, or animals. Humans share 7,000 genes with a worm named C. elegans! The main difference between organisms lies in their total number of genes, how the genes are arranged, and which ones are turned on or off in different cells. A gene from Arctic flounder that keeps the fish from freezing was introduced into strawberries to extend their growing season in northern climates.

Genetic engineering has been used with plants to make:

1. fruits or vegetables that ripen differently;
2. plants that are resistant to disease, pests, selected herbicides, or environmental conditions (such as drought); and
3. plant foods with desirable nutritional characteristics

Using the same principle of gene transfer that gives plants more desirable traits, scientists have started working with GM animals, also called transgenic animals. Animal biotechnology research has largely focused on producing transgenic animals for the study of human disease or to produce drugs. There are many more barriers to using genetic engineering in the breeding and production of animal foods than with plant foods, such as difficulties working with living tissues and life cycles. Currently seeking Food and Drugs Administration approval is a variety of Atlantic salmon that grows to market weight in about 18 months, compared to the 24 to 30 months that it normally takes for a fish to reach that size.

Bioengineered foods are regulated by the US Food and Drug Administration, the US Department of Agriculture, and the Environmental Protection Agency.

Pros and cons revolve around environmental, health, agricultural, nutritional, ethical and moral concerns.

NUTRITION WEB EXPLORER

National Heart, Lung, and Blood Institute (NHLBI):
http://hp2010.nhlbihin.net/atpiii/calculator.asp
At this website, you can use an interactive tool to assess your 10-year risk for heart disease. What is your risk?

American Cancer Society: www.cancer.org/docroot/PED/PED_0.asp
On the website for the American Cancer Society, click on "Food and Fitness" and then click on "ACS Guidelines for Eating Well and Being Active" or "Cooking Smart." What tips do they give to reduce your cancer risk?

Center for Science in the Public Interest: www.cspinet.org/quiz
To prevent disease, you need to eat a healthy diet. Click on "Rate Your Diet Quiz." How does your diet rate?

American Diabetes Association: www.diabetes.org
On the home page of the American Diabetes Association, click on "Nutrition," then click on "Eating Out." What advice does it give if you go to a fast-food restaurant?

National Restaurant Association: www.restaurant.org/foodsafety
Click on "Healthy Dining Finder." What is this program?

CHAPTER REVIEW QUIZ

Key Terms: Matching

1. Atherosclerosis
2. Risk factor
3. Metabolic syndrome
4. TLC diet
5. Stroke
6. Ischemic stroke
7. Hemorrhagic stroke
8. Hypertension
9. Secondary hypertension
10. Primary hypertension
11. Cancer
12. Carcinogen
13. Cruciferous vegetables
14. Diabetes mellitus
15. Hyperglycemia

a. Cancer-causing substance
b. A stroke due to a ruptured brain artery
c. High levels of blood sugar
d. Damage to brain cells resulting from an interruption of blood flow to the brain
e. A combination of risk factors that greatly increase a person's risk of developing coronary heart disease
f. A disorder of carbohydrate metabolism characterized by high blood sugar levels and inadequate or ineffective insulin
g. Vegetarians who eat fish
h. A group of diseases characterized by unrestrained cell division and growth that can disrupt the normal functioning of an organ
i. Most common type of stroke, in which a blood clot blocks an artery or vessel in the brain
j. A habit, trait, or condition associated with an increased chance of developing a disease
k. High blood pressure
l. Members of the cabbage family containing phytochemicals that may help prevent cancer.
m. A condition characterized by plaque buildup along the artery walls; it is the most common form of arteriosclerosis
n. A form of hypertension whose cause is unknown
o. A low-saturated-fat, low-cholesterol eating plan designed to fight cardiovascular disease and lower LDL
p. Individuals who eat a type of vegetarian diet in which no eggs or dairy products are eaten; their diet relies exclusively on plant foods
q. Persistently elevated blood pressure caused by a medical problem

Multiple Choice

1. Which of the following is not a condition of metabolic syndrome?
    a. Reduced fasting glucose
    b. High blood triglycerides
    c. Elevated blood pressure
    d. Excessive abdominal obesity

2. Which is not a characteristic of the TLC diet?
    a. Less than 7 percent of kcalories from saturated fat
    b. Less than 200 milligrams of dietary cholesterol per day
    c. Less than 2,000 kcalories consumed per day
    d. Both a and c

3. What percent of stroke survivors recover almost completely?
    a. 10%
    b. 25%
    c. 35%
    d. 55%

4. All of the following are lifestyles modifications to prevent hypertension except:
    a. Limit consumption of alcohol to no more than 14 drinks per week
    b. Engage in aerobic physical activity three times per week
    c. Reduce dietary sodium intake to no more than 2400 mg per day
    d. Adopt the DASH eating plan

5. What percent of kcalories does the DASH diet recommend come from fat?
    a. 20 percent
    b. 27 percent
    c. 30 percent
    d. 35 percent

6. All of the following play a role in certain types of cancer except:
    a. Physical inactivity
    b. Body weight
    c. Red meat intake
    d. Water consumption

7. Which of the following are cancer promoters?
    a. Estrogen
    b. Cigarette smoking
    c. Alcohol
    d. Fat
    e. All of the above

8. Which of the following is NOT a cruciferous vegetable?
    a. Mustard greens
    b. Kale
    c. Cabbage
    d. Carrots

9. What symptom is shared by people who have type 1 diabetes and type 2 diabetes?
    a. Hypoglycemia
    b. Hyperglycemia
    c. Must inject insulin in order to have enough insulin in the body
    d. Are typically overweight and inactive, causing the disease to occur

10. According to the Exchange List, ½ cup of cooked carrots is considered:
    a. 1 non-starchy vegetable
    b. 1 starchy vegetable
    c. 1 starch
    d. 1 other carbohydrate

True/False

1. Coronary heart disease is the most common form of CVD.
    a. True   b. False

2. Most heart attacks are caused by a clot in a coronary vein at the site of narrowing and hardening that stops the blood flow.
    a. True   b. False

3. Because blood clots play a major role in causing strokes, drugs that inhibit blood coagulation may prevent clot formation.
    a. True   b. False

4. Ischemic strokes occur because a blood vessel in the brain ruptures and causes bleeding in the surrounding brain tissue.
    a. True   b. False

5. The brain uses about 50 percent of the body's oxygen and 25 percent of the glucose circulating in the blood.
    a. True   b. False

6. Strokes and heart disease have some of the same controllable risk factors, including high blood pressure, diabetes, and obesity.
    a. True   b. False

7. Hypertension is known as the "silent killer" because it has no early warning signs.
    a. True   b. False

8. Myocardial infarction is the term used to refer to plaque buildup along artery walls.
   a. True   b. False

9. Heart disease is the number one leading cause of death in the United States
   a. True   b. False

10. Most adults with diabetes have blood pressure higher than 120/80 mm Hg.
   a. True   b. False

Short Answer

1. What occurs if the artery takes blood to the brain?

   a stroke

2. What is the difference between primary and secondary hypertension?

   P- has no casp
   S- cased by evelated blood presure

3. What do systolic pressure and diastolic pressure measure?

   s- top # measures blood w/in artes when the hearts pmps
   d- pressure in the arties when the heart is resting between beats

4. What substance in fruits and vegetables fights cancer formation?

   phytochemicals

5. List three complications of diabetes mellitus.

   blindness, dental disease, nervous System disease

STUDENT WORKSHEET 11-1

TOPIC: Diabetic Diet

1. Use the following website to plan a dinner that includes: 2 lean meat exchanges, 1 reduced fat milk exchange, 1 starch exchange, 1 fruit exchange, 1 nonstarchy vegetable exchange, and 2 fat exchanges. Write out the menu on this sheet.

http://www.dietsite.com/dt/Diets/Diabetes/exchange_system.asp

2. How many kcalories does this meal include? How many grams of carbohydrate, fat, and protein does this meal include?

STUDENT WORKSHEET 11-2

TOPIC: Heart-Healthy Desserts

You have been asked to make six heart-healthy desserts for an American Heart Association event. Find six recipes for a variety of desserts. Explain why you chose each one.

STUDENT WORKSHEET 11-3

TOPIC: Anti-Cancer Diet

You have been asked to make a healthy box lunch that follows the American Cancer Society guidelines. The lunch must include an entrée, two sides, a dessert, and a beverage—no hot items. Your box lunch will be graded on overall appearance, balance of shape and textures, flavors, acceptability of combinations, and how well it meets and ACS guidelines.

Entrée:

Side:

Side:

Dessert:

Beverage:

CHAPTER 12 WEIGHT MANAGEMENT AND EXERCISE

LEARNING OBJECTIVES

Upon completion of the chapter, the student should be able to:

1. Define obesity and overweight
2. List one advantage and one disadvantage of each of the three methods of measuring obesity
3. List the health implications of obesity
4. Explain possible causes of obesity
5. List the six components of a comprehensive treatment program for obesity
6. Describe basic concepts of nutrition education to consider in planning weight-reducing eating plans
7. List five benefits of exercise
8. Explain how behavior and attitude modification can be used to help a person lose weight
9. Explain when drugs and surgery may be used to treat obesity
10. Give an advantage and a disadvantage of using drugs to lose weight
11. Identify strategies that appear to support weight maintenance
12. Evaluate diet books
13. Identify nutrient needs for athletes and plan menus for athletes

CHAPTER OUTLINE

1. **Introduction**

   As of 2004, 66% of American adults are overweight or obese. The prevalence of childhood overweight and obesity has doubled in the last 20 years.

   As the prevalence of overweight and obesity has increased, so have related health care costs.

   Obesity is a disease, not a moral failing, in which many factors are involved: social, behavioral, cultural, physiological, metabolic, and genetic.

2. **How to measure weight/obesity**

   Ways to measure obesity:

   Body mass index (BMI)—calculated by dividing body weight (in kilograms) by height (in meters) squared, BMI is a more sensitive indicator and a better measure of the amount of fat you have compared with height/weight tables. A BMI of 25–29.9 is considered overweight. A BMI of 30 or more is considered obese.

   Measure abdominal fat—abdominal fat is more health-threatening than fat in the hips or thighs. If an individual has a waist greater than 40 inches for man or 35 inches for women, he/she is at an increased risk of developing serious health problems.

   Percentage of body fat—for men under 25% is desirable, for women under 35%.

3. **Health implications of obesity**

   An obese individual is at increased risk for:

   - Type 2 diabetes
   - High blood cholesterol levels
   - Hypertension
   - Cardiovascular disease
   - Stroke
   - Sleep apnea and respiratory problems
   - Certain types of cancer
   - Gallbladder disease
   - Osteoarthritis
   - Complications in pregnancy and childbirth

   Conditions aggravated by obesity include arthritis, varicose veins, and gallbladder disease.

   Obesity creates a psychological burden.

   Losing weight improves health.

4. **Theories of obesity**

   Obesity is caused by an interaction of:

   - Genetics, heredity
   - Environmental influences—related to food intake, food availability, and physical activity
   - Metabolic—almost limitless capacity of the body to store fat
   - Behavioral—psychological and emotional

5. **Treatment of obesity**

   The goals of most weight-loss programs have focused on short-term weight loss—critics feel this is not a valid measure of success because it is not associated with health benefits—critics also point out that weight-loss goals are set too high when even a small weight loss can improve many problems associated with overweight such as:

   - Lower blood pressure and thereby the risks of high blood pressure
   - Reduce abnormally high levels of blood glucose associated with diabetes
   - Bring down blood levels of cholesterol and triglycerides associated with cardiovascular disease
   - Reduce sleep apnea or irregular breathing during sleep
   - Decrease the risk of osteoarthritis in the weight-bearing joints
   - Decrease depression
   - Increase self-esteem

   One approach to obesity is the nondieting approach in which obese people are taught to eat fewer fat kcalories and to get regular exercise.

Discuss the components of a comprehensive approach to treating obesity.

Diet and nutrition education

1. Kcalories should not be overly restricted.

2. Fat should be restricted to about 30% or less, protein to 15%, and carbohydrate to 55% or more.

3. Emphasize fruits, vegetables, whole grains, and fat-free or low-fat milk and milk products. Include lean meats, poultry, fish, beans, eggs, and nuts. Watch intake of saturated fats, trans fats, cholesterol, salt, and added sugars.

4. No foods should be forbidden.

5. Eat regular meals and snacks.

6. Portion control is vital.

7. Variety, balance, and moderation are crucial.

Fad diets may help you take off some pounds initially, but not necessarily in the long term.

While losing weight, regular monitoring of your weight will be essential to help you maintain your lower weight.

Exercise *helps lose pounds and keep them off. A consistent pattern of exercise is important. Exercise must be started slowly, with enjoyment and commitment as major goals. Discuss benefits of exercise such as improved functioning of cardiovascular system; and increased ability to cope with stress, anxiety, and depression;, and improved self-image.*

Behavior and attitude modification: *Behavior modification deals with identifying and changing behaviors that affect weight gain. Setting small, measurable and realistic goals (called shaping) help with successful weight loss.*

Categories of behavior and attitude modification:

- *Self-monitoring (such as a food diary)*
- *Stimulus or cue control (such as storing food out of sight)*
- *Eating behaviors (such as eating only at the kitchen table)*
- *Reinforcement or self-reward (such as calling a friend)*
- *Self-control (such as using relaxation techniques or positive self-talk to respond to cravings)*
- *Attitude modification (never say never)*

Social support: *It's important when families and friends are supportive.*

Maintenance support: *Strategies that support weight maintenance include working out a realistic eating plan, learning skills for dealing with high-risk situations, continued self-monitoring, exercise, social support, use of other strategies that worked during weight loss, and dealing with unrealistic expectations about being thin.*

Drugs: *Drugs approved by the FDA for long-term use may be used for obese people and people with a BMI of 27 or more who also have diseases such as diabetes. All the prescription weight-loss drugs work by suppressing the appetite. Xenical is a lipase* *inhibitor that is available over-the-counter under the name Alli. Herbal preparations are not recommended.*

Surgery: *Weight-loss surgery is an option for people with clinically severe obesity—defined as a BMI of 40 or higher or 35 or higher with other risk factors. This treatment is used only for people in whom other methods have failed. Weight-loss surgery provides medically significant sustained weight loss for more than 5 years in most people. Obesity surgery alters the digestive process— reduce stomach size(adjustable gastric banding) or cause malabsorption (gastric bypass).*

6. **Menu planning for weight loss and maintenance**

7. **Underweight**
   Underweight = 10 percent below desirable body weight
   Tips to increase kcalories: add cheese to sandwiches and entrees, eat small meals frequently, drink beverages with kcalories, use margarine and other fats liberally, add skim milk powder to soups, etc.

8. **Nutrition for the athlete**
   Many athletes require between 3,000 to 6,000 kcalories daily.

   Carbohydrates and fat are the primary fuel sources for exercise. An appropriate diet for many athletes consists of 60 to 65 percent of kcalories as carbohydrate, 30 percent or less as fat, and enough protein to provide between 1 to 1.5 grams per kilogram body weight.

   Water is the most crucial nutrient for athletes. They need about 1 liter for every 1,000 kcalories consumed. For moderate exercise without extreme temperatures or duration, cold water is the best choice. Athletes need 2 cups of water for every pound lost during exercise.

   Carbohydrate or glycogen loading is a regimen involving both decreased exercise and increased consumption of carbohydrates before an event to increase the amount of glycogen stores.

   Menu-planning guidelines for athletes:
   - *Offer a variety of foods.*
   - *Offer good sources of complex carbohydrates.*
   - *Don't offer too much protein and fat.*
   - *Offer a variety of fluids.*
   - *Make sure iodized salt is on the table.*
   - *Be sure to include sources of iron, calcium, and zinc at each meal.*
   - *The most important meal is the precompetition meal. It should consist of mostly complex carbos and be moderate in protein and low in fat. It should include several cups of fluid. Smaller meals can be served 2–3 hours before the competition, larger meals 3–4 hours before.*
   - *After competition and workouts, again emphasize complex carbohydrates to ensure glycogen restoration.*

9. **Food Facts: Sports Drinks**
   Sports drinks contain a dilute mixture of carbohydrate and electrolytes. They are purposely made to be weak solutions so they can empty faster from the stomach. During exercise lasting 60 minutes or more, sports drinks can help replace water and electrolytes and provide some carbohydrates for energy. Most sports drinks contain several carbohydrate sources: such as glucose, fructose, and sucrose. Carbohydrate concentrations over 8 percent can cause stomach cramps and diarrhea.

   Energy drinks contain large doses of caffeine and other legal stimulants like ginseng.

10. **Hot Topic: Fad Diets**
    There is little scientific research to corroborate the theories expounded in the majority of diet books on the market. Several diets are examined.

NUTRITION WEB EXPLORER

National Institute of Diabetes and Digestive and Kidney Diseases: www.niddk.nih.gov
On the home page for the NIDDK, under "Health and Disease Topics," click on "Weight Control." Next, click on "Publication" and choose one to read and write a paragraph about below.

U.S. Department of Agriculture: www.ars.usda.gov/is/AR/archive/mar06/diet0306.htm
Read this article and find out the strongest predictor for weight loss.

Sports Science News: www.sportsci.org
On the home page, click on the index on the left for "Sports Nutrition." Read any one of the articles listed. Summarize in one paragraph what you read.

Shape Up America!: www.shapeup.org/10000steps.html
Read about a program that encourages participants to use a pedometer and walk 10,000 steps a day for physical fitness.

CHAPTER REVIEW QUIZ

Key Terms: Matching

1. Body mass index b
2. Overweight d
3. Obese a
4. Pre-competition meal c

a. Having a body mass index of 30 or greater
b. A method of measuring the degree of obesity that is a more sensitive indicator than height-weight tables
c. The meal closest to the time of a competition or event
d. Having a body mass index of 25 or greater

Multiple Choice

1. As of 2004, what percent of American adults are either overweight or obese?
   a. 35%
   b. 50%
   c. 66%
   d. 75%

2. 300,000 people die each year due to problems attributed to obesity, including:
   a. Osteoarthritis
   b. Gallbladder disease
   c. Hypertension
   d. Both b and c
   e. All of the above

3. Obesity is caused by an interaction of the following:
   a. Environmental factors
   b. Metabolic factors
   c. Behavioral factors
   d. Genetic factors
   e. All of the above

4. Which is not a common side effect or Meridia (sibutramine), which was approved in 1997 by the FDA to reduce the appetite?
   a. Diarrhea
   b. Insomnia
   c. Dry mouth
   d. Headache

5. In order to be eligible for weight-loss surgery, one must be considered severely obese, which is a BMI of:
    a. 30
    b. 35
    c. 40
    d. 45

6. A person is considered underweight if the BMI is less than:
    a. 21
    b. 20
    c. 19.5
    d. 18.5

7. One's fluid loss will depend on:
    a. How large one is
    b. One's fitness level
    c. Genetics
    d. Both b and c
    e. All of the above

8. How much protein do endurance athletes require?
    a. 0.8 grams per kilogram
    b. 1.0 to 1.2 grams per kilogram
    c. 1.2 to 1.6 grams per kilogram
    d. 1.6 to 1.7 grams per kilogram

9. How long does it take to fill glycogen stores without interruption from another practice or competition?
    a. 3 hours
    b. 10 hours
    c. 20 hours
    d. 24 hours

10. What is the most crucial nutrient for athletes?
    a. Kcalories
    b. Water
    c. Sodium
    d. Potassium

True/False

1. The National Institute of Health defines being overweight as a BMI over 30.
   a. True
   b. False

2. Having excessive abdominal fat is more health-threatening than is fat in the hips or thighs.
   a. True
   b. False

3. When kcalorie intake exceeds expenditures, each fat cell can balloon to more than 8 times its original size.
   a. True
   b. False

4. Healthy eating plans contain fat-free or low-fat milk products.
   a. True
   b. False

5. Aerobic capacity improves when exercise increases the heart rate to a target zone of 60 to 75 percent of the maximum heart rate.
   a. True
   b. False

6. Xenical (Orlistat) is a lipase inhibitor that interferes with lipase function, decreasing dietary fat absorption by 30 percent.
   a. True
   b. False

7. Self-monitoring, stimulus control, and eating behaviors are considered metabolic modifications.
   a. True
   b. False

8. To sustain weight loss in adulthood, one should participate in 30 to 60 minutes of moderate daily physical activity.
   a. True
   b. False

9. Most weight loss drugs work by decreasing the appetite or fat absorption.
   a. True
   b. False

10. Decreasing glycogen stores by up to 50 to 80 percent will enhance performance by providing more energy during lengthy competition.
    a. True
    b. False

Short Answer

1. What is BMI, and how does it apply to men and women?

   Body mass index is a method of measuring the degree of obesity. It is a direct calculation based on weight + height

2. What are three benefits of exercise?

   ↑ self image, ↑ stamina, + ↑ resistance to fatigue

3. How can one calculate the maximum heart rate during exercise?

   subtract your age from 220

4. Which foods should be forbidden when one is attempting to lose weight?

   None

5. How does gastric bypass surgery result in severe weight loss?

   is a small pouch created to resist food intake. a small section of the small intestine is attached to the pouch to allow food to bypass the lower stomach + part of the small upper intestine. this reduces amounts of kcals + nutrients absorbed by the body

STUDENT WORKSHEET 12-1

TOPIC: Weight Loss Diet

If you eat 500 fewer kcalories each day for one week, you will lose about 1 pound. Write down below everything you ate yesterday, including estimated portion size. Next, decide which items you could delete from your diet or simply eat less of. Find the amount of kcalories in each food or beverage you would cut back on, and then add up the total kcalories. Were you able to actually cut back 500 kcalories? Would you really be able to give up these foods/beverages?

Foods/Beverages	Portion Size	Food/Beverage & Amount to Omit	Kcalories in Food/ Beverage Omitted
			Total Kcal:

STUDENT WORKSHEET 12-2

TOPIC: Pre-competition Meal

You have been asked to suggest a precompetition meal for a men's basketball team. The game is at 4:00 PM and the meal will be served at 12:30 PM. Write your menu below and be sure to include beverages. Allow for the players to have some food choices. When your menu is complete, do a nutritional analysis of a possible meal to determine if you kept the fat at or below 20%.

MENU ITEMS	PORTION SIZE

Meal Used to Do Nutrient Analysis: ______________________________

____

____

Percent Fat in Meal: ______________

CHAPTER 13 NUTRITION OVER THE LIFE CYCLE

LEARNING OBJECTIVES

Upon completion of the chapter, the student should be able to:

1. Explain the benefits of good nutrition to mother and baby during pregnancy
2. Identify nutrients of special concern during pregnancy and their food sources
3. Explain the possible effects of alcohol, fish, caffeine, and artificial sweeteners during pregnancy
4. Plan menus for women during pregnancy and lactation
5. Describe what an infant should be fed during the first year, including the progression of solid foods
6. Give five reasons why breast-feeding is preferable to bottle feeding
7. Describe how to ensure enjoyable mealtimes with young children and teach them good eating habits
8. Plan menus for preschool and school-age children
9. Identify the nutrients that children and adolescents are most likely to be lacking and their food sources
10. Describe influences on children's and adolescents' eating habits
11. Plan menus for adolescents
12. Distinguish among anorexia nervosa, bulimia nervosa, binge eating disorder, and female athlete triad
13. Describe factors that influence the nutrition status of older adults
14. Identify nutrients of concern for older adults and their food sources
15. Plan menus for healthy older adults
16. Describe ways to prevent the development of obesity during childhood

CHAPTER OUTLINE

1. **Nutrition during pregnancy**

   Discuss embryo, fetus (from 8 weeks), amniotic sac, placenta.

   The first trimester is the critical period of cell differentiation during which the growing baby is most susceptible to damage from nutritional deficiencies and alcohol.

   The nutritional status of women before and during pregnancy influences both the mother's and the baby's health. Weight gain is about 25–35 pounds for normal weight women.

   Factors that place a woman at nutrition risk during pregnancy include an inadequate diet, underweight or overweight at time of conception, smoking, and gaining too few or too many pounds during pregnancy.

   A newborn's weight is the #1 indicator of his or her future health status. A newborn who weighs less than 5½ pounds is a low-birth-weight baby.

   During pregnancy, a woman should eat 340 kcalories more during the second trimester and an additional 452 kcalories during the third trimester. She must also pack more protein and nutrients into those extra kcalories.

During the first 13 weeks of pregnancy (first trimester), total weight gain is 2–4 pounds; thereafter about 1 pound a week is normal.

Pregnancy is no time to diet, but it's fine to exercise appropriately.

Protein needs increase 25 grams.

Requirements for the essential fatty acids, linoleic and alpha-linolenic acids, increase—important for growth and development of the brain.

Calcium, vitamin D, phosphorus, and magnesium are necessary for the proper development of the skeleton and teeth. Calcium may help reduce pregnancy-induced hypertension and edema.

Need for folate (to make new cells) and iron (to make RBCs) increase significantly during pregnancy. Folate is critical in the first 4–6 weeks of pregnancy because this is when the neural tube, the tissue that develops into the brain and spinal cord, forms. Insufficient folate can cause birth defects of the brain and spinal cord. Folate is also critical during the entire pregnancy.

Need for vitamin B12 also increases as it works with folate to make new cells.

Iron supplements are necessary during 2nd and 3rd trimesters of pregnancy because a woman can't meet increased needs through diet alone. Iron is needed for increase in maternal blood volume and development of fetal blood supply.

Sodium restriction is unnecessary during pregnancy.

2. **Diet-related concerns during pregnancy**
   To plan menus, certain diet-related concerns must be addressed.

   Nausea and vomiting: Nausea and vomiting can occur during the first 17 weeks of pregnancy (although it can last all the way through pregnancy). Dietary advice in the past has concentrated on small, carbohydrate-rich meals and tea and crackers. For many women, this dietary advice doesn't work. Recent advice says to eat whatever food you feel you can keep down, even when it means a food that isn't terribly nutritious such as potato chips. The American College of Obstetricians and Gynecologists recommends taking a multivitamin at the time of conception and taking vitamin B6 to possible decrease symptoms.

   Cravings and aversions

   Constipation—can be counteracted by eating more high-fiber foods, drinking more fluids, and getting more exercise.

   Heartburn—can be counteracted by eating small and frequent meals, eating slowly and in a relaxed atmosphere, avoiding caffeine, wearing comfortable clothes, and not laying down after eating.

   Alcohol and pregnancy don't mix: Alcohol crosses placenta and can limit amount of oxygen delivered to fetus. Heavy consumption of alcohol may cause fetal alcohol syndrome. Even moderate drinkers have babies with mild symptoms of fetal alcohol syndrome.

Pregnant women should not eat shark, swordfish, king mackerel, or tilefish. These long-lived fish contain the highest levels of methylmercury. Eat up to 12 ounces a week of a variety of fish and shellfish that are lower in mercury. Five of the most commonly eaten fish that are low in mercury include shrimp, canned light tuna, salmon, Pollock, and catfish. Albacore tuna has more mercury than canned light tuna.

Caffeine should be used in moderation. Artificial sweeteners are safe.

Pregnant women can follow the Daily Food Guide for Pregnancy (Page 469).

Menu-planning guidelines for pregnant and lactating women:

1. *Offer a varied and balanced selection of nutrient-dense foods. Because energy needs increase less than nutrient needs, empty kcalories are rarely an acceptable choice.*
2. *In addition to traditional meat entrees, have some entrees based on legumes and/or grains and dairy products. Beans, peas, rice, pasta, and cheese can be used in many entrees.*
3. *Be sure to offer dairy products made with fat-free or low-fat milk.*
4. *Use a variety of whole-grain and enriched breads, rolls, cereals, rice, pasta, and other grains in menus.*
5. *Use assorted fruits and vegetables in all areas of the menu, including appetizers, salads, entrees, side dishes, and desserts.*
6. *Be sure to have good sources of problem nutrients: the essential fatty acids, calcium, vitamin D, magnesium, folate, vitamin B12 and iron.*
7. *Be sure to use iodized salt.*

3. **Nutrition and menu planning during lactation**

Lactating mothers normally produce about 25 ounces of milk a day.

During lactation, 330 additional kcalories are needed during the first 6 months, which increases to 400 during the 7th to 12th month. More protein is also needed, as well as a number of vitamins and minerals. Also at least 3 to 4 quarts of water are needed each day to prevent dehydration.

Nutritional deficiencies are more likely to affect the quantity of milk the mother makes, rather than the quality.

A balanced, varied and adequate diet of at least 1,800 kcalories per day is critical to successful breast-feeding and infant health. Iron supplementation is recommended.

Occasional consumption of alcohol will probably have no consequences. Moderate use of caffeine is okay.

Menu-planning guidelines are the same as those for pregnant women, with an emphasis on fluids, dairy products, fruits, and vegetables.

4. **Nutrition during infancy**

Infants generally double their birth weight in the first 4 to 5 months and then triple their birth weight by the first birthday. An infant will also grow 50 percent in length by the first birthday.

Newborns need a plentiful supply of all nutrients. The DRI is set for infants from 0 to 6 months, and then from 7 to 12 months.

For the first 4 to 6 months of life, the source of all nutrients is breast milk or formula. Breast milk is recommended for all infants under ordinary circumstances from birth to 12 months.

Advantages of breast-feeding:

- *Breast milk is nutritionally superior to any formula.*
- *Newborns are less apt to be allergic.*
- *Suckling promotes the development of the infant's jaw and teeth.*
- *Breastfeeding promotes a close relationship.*
- *Breast milk is less likely to be mishandled.*
- *Breast milk helps the infant build up immunities to infectious disease because it contains the mother's antibodies to disease.*
- *Breast-feeding may reduce the risk of breast cancer for the mother.*
- *Breast-feeding is less expensive.*
- *Breast-fed infants have lower rates of hospital admission, ear infections, diarrhea, rashes, allergies, and other medical problems*

Vitamin D supplements are recommended for breast-fed infants beginning at 2 months of age and until they begin taking at least 17 fluid ounces of vitamin D-fortified milk daily.

Formula feeding is an acceptable substitute for breastfeeding and has some advantages (such as convenience). All formulas must meet nutrient standards set by the American Academy of Pediatrics. Cow-based formulas are normally used unless the baby is allergic to the protein or sugar in milk. The Food and Drug Administration approved the addition of 2 fatty acids to infant formula: DHA and AA, which are both present in breast milk and are thought to enhance the mental and visual development of infants.

Whether breast-fed or formula fed, the infant's iron stores are low by 4 to 6 months, at which time they typically start to eat iron-fortified cereals.

5. **Feeding the infant**

Ideally the feeding schedule should be based on reasonable self-regulation by the baby.

Formula-fed babies are fed about every 4 hours, breast-fed babies about every 2 to 3 hours.

The mother has to breastfeed the child as soon as possible after delivery to enhance success. Colostrum, a yellowish fluid, is the first secretion to come from the breast a day or so after delivery. It is rich in proteins, antibodies, and other factors that protect against infectious disease. Colostrum changes to transitional milk between the third and sixth days, and then by the tenth day the major changes are finished.

The breast-feeding process begins with the infant using a sucking action that stimulates hormones to move milk into the ducts of the breast-milk letdown.

Babies are ready to eat semi-solid foods when they can sit up and open their mouths. This usually occurs between 5 to 7 months of age. Other signs include:

- Doubled birth weight
- Drinking more than a quart of formula a day
- Baby seems hungry often
- Baby opens mouth in response to seeing food coming

Eating solid food involves a number of difficult steps. The gag reflex prevents choking.

Order of foods

4–7 months:	*Iron-fortified baby cereals*
	Pureed then textured vegetables
	Pureed then textured fruit
	Fruit juice (start at 6 months, dilute at first)
8–11 months:	*Mashed or diced soft fruit*
	Mashed or soft cooked vegetables
	Finely cut meat/poultry
	Mashed cooked beans or peas
	Cottage cheese, yogurt, or cheese strips
	Pieces of soft bread
	Crackers
12 months:	*Cut-up table foods*
	Whole milk
	Whole eggs

Review baby bottle tooth decay, palmar grasp (from about 6 months), and pincer grasp (from about 8 months). More than a half cup of juice daily can result in growth failure if substituted for breast milk or formula.

Several foods should be avoided during the first year. Because honey and liquid corn syrup may be contaminated with botulism, these foods may cause food poisoning or foodborne illness in children younger than 1 year.

Certain foods also are more apt than others to cause choking. They include nuts, seeds, raisins, hot dogs, popcorn, hard candies, whole grapes, peanut butter, chunks of apple or pear, celery, cherries with pits, large chunks of food, and other raw vegetables or fruits.

Other foods also are more apt to cause allergies: milk, eggs, wheat, nuts, chocolate, and shellfish. Whole milk and eggs are usually introduced at about 10 to 12 months.

6. **Nutrition during childhood**

Around the age of one year, the growth rate decreases markedly. Yearly weight gain now approximates 4 to 6 pounds and children will grow about 3 inches per year up to age 7 and then 2 inches per year until puberty. Until adolescence, growth will come in spurts.

After 1 year of age, children start to lose baby fat and become leaner and legs become longer.

By 2 years of age the baby teeth are almost all in and between ages 6 and 12 these teeth are replaced with permanent teeth.

Energy needs of children of similar age, sex, and size can vary due to differing BMRs, growth rates, and activity levels.

Both energy and protein needs decline gradually per pound of body weight. A decreased appetite in childhood is normal. During growth spurts, appetite increases and the requirements for kcalories and nutrients are greatly increased.

Although kcalorie and protein intakes are rarely inadequate, there are concerns about iron intake. Lack of iron can cause decreased energy and affect behavior, mood, and attention span.

Children and adolescents from 4 to 18 years of age should keep fat intake between 25 to 35 percent of total kcalories, with most fats coming from monounsaturated and polyunsaturated fatty acids. Saturated and trans fats should be limited to 10% or less of total kcalories.

Preschoolers normally exhibit food jags and can be finicky about how food is prepared and served. Toddlers (ages 1 to 3) tend to be pickier eaters than older preschoolers (ages 4 to 5).

Tactics for dealing with preschoolers' (and school-age children's) food habits:

1. *Make mealtime as relaxing and enjoyable as possible.*
2. *Don't nag, bribe, force, or even cajole a child to eat. Be calm.*
3. *Allow children to choose what they will eat from two or more healthy choices.*
4. *Let children participate in food selection and preparation.*
5. *Respect your child's preferences when planning meals, but don't make your child a quick peanut butter sandwich if he or she rejects your dinner.*
6. *Make sure your child has appropriately sized utensils and can reach the table comfortably.*
7. *Preschoolers love rituals, so start them early with the habit of eating three meals plus snacks each day at fairly regular times.*
8. *Expect preschoolers to reject new foods at least once, if not many times.*
9. *Let the child serve himself or herself small portions.*
10. *Do not use desserts as a reward for eating meals.*
11. *Ask children to try new foods and praise them when they do*
12. *Be a good role model.*
13. *Be consistent at mealtimes.*
14. *If all else fails, keep in mind that children under six have more taste buds.*

The school-age child is a much better eater.

Children's eating habits are influenced by family, friends, teachers, availability of school breakfast and lunch programs, and television.

Studies show that the more television watched, the greater the incidence of obesity. Both obesity and inactivity are on the rise in school-age children, especially adolescents.

Parents can start their children on a nutritious eating path by being a good role model, having nutritious food choices available at home, serving regular meals, limiting television watching, and encouraging physical activity.

7. **Menu planning for children**
   By age 4, children can eat amounts that count as regular servings eaten by older family members. Children 2 to 3 need the same variety of foods as 4 to 6-year olds but may need smaller portions (about 2/3). 2 to 6-year olds need 2 full servings from the milk group each day.

   Menu planning guidelines for preschoolers include offering simply prepared foods, colorful foods, raw vegetables, some soft foods and some chewy foods, lightly seasoned foods, carbohydrate foods, smooth-textured foods, bite-sized foods, warm (not hot) foods, cut-up fruits and vegetables, and foods that they can't choke on.

   Children learn to like new foods by being presented with them repeatedly—as many as 12 or more times. Put a small amount of a new food along with a meal and don't require the children to eat it if he/she doesn't want to.

   Menu planning guidelines for school-age children include serving a wide variety of foods (making sure to include children's favorites), nutritious snacks, moderate amounts of fat, and good sources of iron, potassium, vitamin E, and fiber.

8. **Nutrition and menu planning for adolescents**
   Adolescence lasts from 11 to 21 years of age. Puberty starts at about age 10 or 11 for girls and 12 or 13 for boys. The growth spurt is intense for 2 to 2-1/2 years, and then there are a few more years of growth at a slower pace. During the five to seven years of pubertal development, the adolescent gains about 20 percent of adult height and 50 percent of adult weight. Whereas the proportion of fat and muscle was similar in males and females before puberty, now males put on twice as much muscle as females, and females gain proportionately more fat. Males also experience a greater increase in bone mass than females.

   Males now need more kcalories, protein, calcium, iron, and zinc for muscle and bone development than females; however, females need increased iron due to the onset of menstruation. Calcium is very important for males and females because half of their peak bone mass is built during this time.

   Teenage boys are more likely to get sufficient nutrients than females. Teenagers make most of their own food choices, which are influenced by their body image, peers, and the media.

   Menu planning guidelines for adolescents include emphasizing complex carbohydrates; offering well-trimmed lean beef, poultry, and fish; offer low-fat and fat-free milk; have nutritious snack choices; emphasize quick and nutritious breakfasts; and offer foods that contain the nutrient most often lacking in their diets: fiber, vitamin E, calcium, potassium, and magnesium.

9. **Eating disorders**
   Understand the characteristics of anorexia nervosa, bulimia nervosa, binge eating disorder, and female athlete triad (disordered eating, no menstruation, osteoporosis).

   The sooner a disorder is diagnosed, the better the chances that treatment can work.

   Treatment usually includes individual psychotherapy, family therapy, cognitive-behavior therapy, medical nutrition therapy, and possibly medications.

10. **Nutrition for the elderly**

The older population (65 or older) numbers about 1 in 8 Americans. The older population will continue to grow significantly in the future.

The maximum efficiency of many organ systems occurs between 20 and 35. After age 35, the functional capability of almost every organ system declines.

BMR declines as we age. We lose muscle mass.

The functioning of the cardiovascular system declines with age. Blood pressure increases. Pulmonary capacity decreases. Kidney function deteriorates.

Factors affecting nutrition status:

Physiological

- Disease
- Less muscle mass
- Activity levels
- Dentition
- Functional disabilities
- Decreased sensitivity to taste and smell
- Changes in GI tract (slowing down, heartburn)
- Medications
- Diminished sense of thirst

Psychosocial

- Cognitive functioning
- Social support

Socioeconomic

- Education
- Income
- Living arrangements
- Availability of federally funded meals

Nutrients of concern to the elderly may include:

Water: Due to decreased thirst sensation and other factors, fluid intake is important.
Vitamin B12: Less is absorbed. Vitamin B12 deficiency can cause a type of anemia as well as nervous system problems.
Folate
Vitamin D: Because milk is the only dairy product with vitamin D added to it, milk is important to getting enough vitamin D, unless an individual has good daily sun exposure. Vitamin D-fortified cereals are also a good source.
Calcium: Current intakes are below the DRI of 1,200 mg for people over 65 years of age.
Zinc: Zinc is important for taste, immune system, and wound healing.

11. **Menu planning for older adults**
    Use the modified Food Guide Pyramid for people over 70 (page 498).

    Explain tips
    1. *Offer moderately sized meals.*
    2. *Emphasize high-fiber foods such as fruits, vegetables, grains, and beans.*
    3. *Moderate the use of fat.*
    4. *Dairy products are important sources of calcium, vitamin D, protein, potassium, vitamin B12, and riboflavin.*
    5. *Offer adequate but not too much protein.*
    6. *Moderate the use of salt.*
    7. *Use herbs and spices to make foods flavorful.*
    8. *Offer a variety of foods, including traditional menu items and cooking from other countries and U.S. regions.*
    9. *Fluid intake is critical so offer a variety of beverages.*
    10. *If chewing is a problem, provide softer foods such as ground meats, cooked beans and peas, mashed potatoes, soft fruits, and soft breads and rolls.*

12. **Food Facts: Food Allergies**
    A food allergy involves an abnormal immune system response. If the response doesn't involve the immune system, it is a food intolerance.

    Symptoms of food allergy may include hives, rashes, stomach cramps, vomiting, diarrhea, wheezing, swelling of the lips or tongue, and itching lips.

    The greatest danger in food allergy comes from anaphylaxis.

    The most common foods that cause allergies are peanuts, tree nuts, shellfish, milk, eggs, wheat, and fish.

13. **Hot Topic: Childhood Obesity**
    Obesity is a serious health problem for children and adolescents.

    The prevalence of obesity in U.S. children has increased from about 5 percent in 1963–1970 to 17 percent in 2003 to 2004. Data from a national nutrition survey comparing data from 1971 to 1974) to date from 2003 to 2004 show increases in overweight among all age groups:

    Among preschool-aged children, aged 2–5 years, the prevalence of overweight increased from 5.0 percent to 13.9 percent.

    Among school-aged children, aged 6–11 years, the prevalence of overweight increased from 4.0 percent to 18.8 percent.

    Among school-aged adolescents, aged 12–19 years, the prevalence of overweight increased from 6.1 percent to 17.4 percent.

    Contributing factors include increased energy intake due to large portion sizes, eating meals away from home, frequent snacking and consuming beverages with added sugar; less physical activity and more sedentary habits; and factors within the home, child care, school, and community environments.

Childhood overweight is associated with various health-related consequences as well as social consequences.
It is important for parents and schools to help children maintain a healthy weight by balancing the kcaloires the child eats with the kcalories they uses through physical activity. The goal for overweight children and teens is to reduce the rate of weight gain while allowing normal growth and development.

NUTRITION WEB EXPLORER

Medline Plus: www.nlm.nih.gov/medlineplus/pregnancy.html
Click on one of the articles listed under "Nutrition." Read the articles and write a one-paragraph summary of what you read.

Food and Nutrition Services, USDA: www.fns.usda.gov/fns
On this home page, click on "Nutrition Education, then click on "Team Nutrition." Find out what Team Nutrition is and what it does.

National Eating Disorders Association: www.nationaleatingdisorders.org
At the top of this home page, click on "Information and Resources" and read one of the articles. Write a one-paragraph summary of what you read.

Federal Interagency Forum on Aging-Related Statistics: www.agingstats.gov
On the home page, click on "Older Americans 2008: Key Indicators of Well-Being." Then click on "Introduction Section." This Table of Contents lists 31 indicators of well-being for older adults. Write down any indicators that you think are nutrition-related.

FirstGov for Seniors: www.seniors.gov
This portal site of FirstGov includes many topics of interest to seniors. Under "Health for Seniors," click on "Staying Health As A Senior" and then click on "Nutrition and Aging." Find out what it tells seniors about water, fiber, and fat.

CHAPTER REVIEW QUIZ

Key Terms: Matching

1. Fetus
2. Amniotic sac
3. Embryo
4. Placenta
5. Preeclampsia
6. Edema
7. Neural tube
8. Spina bifida
9. Heartburn
10. Fetal Alcohol Syndrome
11. Colostrum
12. Pincer grasp
13. Palmar grasp
14. Anorexia nervosa
15. Bulimia nervosa
16. Female athlete triad

a. A set of symptoms occurring in newborn babies that is due to alcohol use by the mother during pregnancy
b. The name of the fertilized egg from conception to the eighth week
c. The ability of a baby from about six months of age to grab objects with the palm of the hand
d. Swelling due to an abnormal accumulation of fluid in the intercellular spaces.
e. An eating disorder most prevalent in adolescent females who starve themselves
f. A birth defect in which parts of the spinal cord are not fused together properly and so gaps are present where the spinal cord has little or no protection
g. The organ that develops during the first month of pregnancy, which provides for exchange of nutrients and wastes between fetus and mother
h. An eating disorder found among female college athletes in which they have disordered eating, osteoporosis, and no menstruation
i. A yellowish fluid that is the first secretion to come from the mother's breast a day or so after the delivery of a baby
j. Hypertension during pregnancy that can cause serious complications
k. The infant in the mother's uterus from eight weeks after conception until birth
l. A painful burning sensation in the esophagus caused by the acidic stomach contents flowing back into the lower esophagus
m. A baby's ability at about eight months of age to use the thumb and forefinger together to pick things up
n. The protective bag, or sac, that cushions and protects the fetus during pregnancy
o. Eating disorder characterized by destructive pattern of excessive overeating followed by vomiting or other "purging" behaviors to control weight.
p. The embryonic tissue that develops into the brain and spinal cord.

Multiple Choice

1. During which week of pregnancy do the brain, spinal cord, and heart begin to develop?
    a. Week 2
    b. Week 3
    c. Week 4
    d. Week 5

2. How much more protein per day do pregnant women need?
    a. 10 grams
    b. 25 grams
    c. 35 grams
    d. Protein needs do not increase at all.

3. Approximately when does a baby acquire the skills to eat solid food?
    a. 12 weeks
    b. 16 weeks
    c. 20 weeks
    d. 24 weeks

4. Which foods are considered safe for infants and toddlers to eat?
    a. Grapes and cherry tomatoes cut in quarters
    b. Popcorn
    c. Raisins
    d. Hot dogs

5. The following is not a food that is apt to cause allergies:
    a. Chocolate
    b. Wheat
    c. Nuts
    d. Rice

6. What are age-appropriate cooking activities for 5-year-olds?
    a. Roll and shape cookies
    b. Peel carrots
    c. Shred cheese or vegetables
    d. Measure ingredients

7. What are appropriate cooking activities for 3-year-olds?
    a. Wash and tear lettuce for salad
    b. Place toppings on pizza or snacks
    c. Slice soft foods with a table knife
    d. Set the table

8. Problem nutrients for children include all of the following except:
    a. Potassium
    b. Fiber
    c. Iron
    d. Calcium

9. How many glasses of milk or milk equivalents must teenagers consume to meet their calcium needs?
    a. 2
    b. 3
    c. 4
    d. 5

10. Which of the following is not a common symptom of anorexia nervosa?
    a. Lanugo
    b. Increased body temperature
    c. Mild anemia
    d. Swollen joints

True/False

1. The lungs begin to form during the sixth week of pregnancy.
    a. True   b. False

2. A low-birth-weight baby is a newborn who weighs less than 6 pounds.
    a. True   b. False

3. The need for folate and iron increases 100 percent during pregnancy.
    a. True   b. False

4. Fruit juice fortified with vitamin C can be consumed by babies who are six months of age.
    a. True   b. False

5. More than a quarter cup of juice daily can result in growth failure if substituted for breast milk or formula.
    a. True   b. False

6. Breast milk is not a good source of vitamin D.
    a. True   b. False

7. An infant should triple in length by its first birthday.
    a. True   b. False

8. For children who are two to three years of age, fat intake should be 30 to 40 percent of total kcalories.
   a. True   b. False

9. Extreme appetite fluctuations during childhood are considered normal.
   a. True   b. False

10. Disordered eating, amenorrhea, and osteoporosis made up the female athlete triad.
   a. True   b. False

Short Answer

1. How much total weight gain is expected during the first 13 weeks of pregnancy?

   2-4 lbs

2. What are three tips for breast-feeding success?

   get early start, avoid artifical nipples, & watch for infections

3. How many calories should a two-year-old consume per day?

   1000

4. After a child's first birthday, how much weight gain is typical each year?

   4-6 lbs per year

5. At what age can a child eat food amounts that count as servings eaten by older family members?

   4-5 years per age

STUDENT WORKSHEET 13-1

TOPIC: Nutritious Snacks for Children/Adolescents

Snacks can make an important contribution to a child's nutritional well-being. Write down below a variety of nutritious snacks that children or adolescents would like.

STUDENT WORKSHEET 13-2

TOPIC: School Lunch Menu Evaluation

Below is an actual school lunch menu for a middle school cafeteria for students from ages 11 to 14. Evaluate the menu using the MyPyramid for Kids and the menu planning guidelines given for school-age children and adolescents. List four evaluative comments below, and then change the menu to incorporate some of your suggestions.

Monday	Tuesday	Wednesday	Thursday	Friday
¼ pound Hotdog on a Bun with fries and sliced carrots OR Chicken Caesar salad with roll with sliced carrots OR Cheese pizza with sliced carrots	Meatball Hoagie with fries and corn OR Turkey Club Sandwich with fries and corn OR Cheese Pizza with fries and corn	Chicken Nuggets with sauce, roll, and broccoli cuts OR Chicken Cobb Salad with roll and broccoli cuts OR Cheese Pizza with broccoli cuts	Pasta with sauce, meatballs, French bread, and tossed salad w/dressing OR Cheese Pizza with tossed salad with dressing	BBQ Pork Sandwich with fries and peas and carrots OR Tuna Salad on Greens with roll and peas and carrots OR Cheese Pizza with peas & carrots
BBQ Chicken Wings with buttered noodles and green beans OR Chicken Caesar Salad with roll and green beans OR Cheese Pizza with green beans	French Toast Sticks with syrup and sausage and sliced carrots OR Chicken Parmesan Sandwich with fries and carrots OR Cheese Pizza with carrots	Cheesesteak Hoagie with fries and mixed veggies OR Chicken Cobb Salad with roll and mixed veggies OR Cheese Pizza with mixed veggies	Fried Chicken with roll, fries, and corn OR Cheeseburger with fries and corn OR Cheese Pizza with corn	Chicken Patty on Bun with celery sticks and cake OR Tuna Salad on Greens with roll, Celery sticks, and cake OR Cheese Pizza with celery sticks and cake

STUDENT WORKSHEET 13-3

TOPIC: Senior Center Menu Evaluation

Below is an actual lunch menu for a senior center. Evaluate the menu using the Modified Food Guide Pyramid for Adults Over 70 and the menu planning guidelines. List four evaluative comments below, and then change the menu to incorporate some of your suggestions.

Monday	Tuesday	Wednesday	Thursday	Friday
Steamed Red Snapper, w/Olive Oil, Chow Mein, Salad, Bread, Dessert	Turkey Pot Pie, corn Bread, Vegetables, Salad, Dessert	Mango and Chicken Salad, Steamed Rice, Vegetables, Bread, Dessert	Beef and Broccoli Stir Fry over Rice, Salad, Bread, Dessert	**$5-$7 Special** Lamb Shanks w/ Assorted Vegetables, Mashed Potatoes, Salad, Bread, Dessert
Baked Ribs Baked Beans and Potatoes, Salad, Vegetables, Bread, Dessert	**St. Patrick Special** Corned Beef and Cabbages, Potatoes, Salad, Bread, Dessert	Chicken Makhani over Rice, Vegetables, Salad, Bread, Dessert	Egg Plant Parmesan, Pasta, Vegetables, Salad, Bread, Dessert	Margarita Shrimp Salad, Steamed Parsley Potatoes, Vegetables, Bread, Dessert
Honey Glazed Salmon, Orzo, Vegetables, Salad, Bread, Dessert	Kahlua Pork, Brown Rice, Salad, Bread, Dessert	Spaghetti with Meat Balls, Garlic Bread, Vegetable, Salad, Dessert	Asian Turkey Salad, Rice Pilaf, Vegetables, Bread, Dessert	Taco Salad, Dessert

ANSWERS TO QUIZZES AND SELECTED STUDENT WORKSHEETS

CHAPTER 1

Matching
1. e
2. o
3. k
4. c
5. m
6. g
7. d
8. l
9. f
10. i
11. n
12. b
13. j
14. p
15. h
16. a

True/False
1. T
2. F
3. T
4. T
5. T
6. T
7. F
8. T
9. T
10. F
11. T
12. F
13. T
14. F
15. F
16. F

Fill in the Blank
1. Triglyceride
2. Protein
3. Calcium
4. Lymph
5. Stomach
6. Bile
7. Minerals
8. Nutrient density
9. Water
10. RDA (Recommended Dietary Allowance)

STUDENT WORKSHEET 1-2

1. Dietary Reference Intake
   Estimated Average Requirement
   Recommended Dietary Allowance
   Adequate Intake
   Tolerable Upper Intake Level
   Estimated Energy Requirement
   Acceptable Macronutrient Distribution Range

2. 1. C
   2. E
   3. B
   4. F
   5. D
   6. A

3. A. varies
   B. RDA
   C. 1,000 mg
   D. AI
   E. girl
   F. 1,000 mg
   G. 130 g carbohydrate, no RDA for fat, 0.8 g/kg protein
   H. 28 grams
   I. There is no AI for total fat for anyone except 0–1 year olds

STUDENT WORKSHEET 1-3

A. 14–18 years old: 65 mg
   Males 19+: 90 mg
   Females 19+: 75 mg
B. RDA
C. 1,000 mg
D. AI
E. Girl
F. 1,000 mg vitamin E
G. 130 mg Carbohydrate, no RDA for fat for adults, 56 g protein for males, 46 gr protein for females
H. 28 g fiber
I. No AI or RDA for total fat for adults

CHAPTER 2

Matching
1. d
2. b
3. e
4. h
5. c
6. a
7. g
8. f

Multiple Choice
1. a
2. c
3. c
4. b
5. d

True/False
1. F
2. T
3. T
4. F
5. T
6. T
7. T
8. F
9. T
10. F

Short Answer
1. Whole fruit contains fiber, which is beneficial for nutrient transport.
2. 1 cup milk, 8 oz yogurt, 1.5 oz of natural cheese, and 2 oz processed cheese.
3. Vitamin D

4. LDL; heart disease
5. Discretionary kcalories
6. Monounsaturated and polyunsaturated fats
7. Saturated and trans fats
8. Aerobic, resistance, and stretching exercises
9. FDA (Food and Drug Administration)
10. Eggs, fish, milk, peanuts, shellfish, soybeans, tree nuts, and wheat

STUDENT WORKSHEET 2-6

1. 9 g fat x 9 kcal.gram = 81 kcal from fat
2. High in saturated fat. Low in cholesterol, sodium, dietary fiber.
3. 2,000 kcal – at bottom of label
4. 4.5 g/fat in 2 cookies
5. Not high in sodium or cholesterol
6. 36 cookies in box
7. 30 grams dietary fiber

CHAPTER 3

Matching

1. f
2. d
3. h
4. c
5. u
6. o
7. l
8. q
9. g
10. r
11. b
12. t
13. a
14. m
15. n
16. k
17. e
18. s
19. i
20. p
21. j

Multiple Choice

1. b
2. c
3. d
4. a
5. d

True/False

1. T
2. T
3. F
4. F
5. T
6. F
7. T
8. T
9. F
10. T

Short Answer

1. Fructose
2. Monosaccharides and disaccharides
3. 100 to 150 grams/day
4. Photosynthesis
5. Soft drinks, candy and sugars, baked goods, fruit drinks, and dairy desserts
6. kidney beans
7. Enzymes
8. One can classify fiber by whether it is soluble in water (soluble and insoluble)
9. 3 servings
10. Dried beans, peas, and lentils

STUDENT WORKSHEET 3-1

1. #3
2. #2
3. #1
4. #3
5. 2 3/4 teaspoons (#1), 2 1/2 teaspoons (#2), 1 1/2 teaspoons (#3)
6. all cereal
7. #2
8. Golden Grahams is #2, Honey Nut Cheerios is #1, and Multi-Grain Cheerios is #3

ANSWERS TO STUDENT WORKSHEET 3-3

Bread #1 is completely refined.
Bread #2 contains whole grains and also refined grains.
Bread #3 is completely whole grain.

CHAPTER 4

Matching

1. j
2. q
3. e
5. h
5. c
6. l
7. g
8. k
9. n
10. p
11. r
12. d
13. f
14. i
15. a
16. o
17. m
18. b

Multiple Choice
1. A
2. D
3. D
4. D
5. A
6. C
7. C
8. C
9. D
10. C

True/False
1. F
2. T
3. T
4. F
5. T
6. F
7. T
8. T
9. F
10. T

Short Answer
1. Saturated fatty acids, monounsaturated fatty acids, and polyunsaturated fatty acids.
2. Point of unsaturation
3. Lecithin
4. HDL travels throughout the body picking up cholesterol, which it brings back to the liver for breakdown and disposal. Therefore, it prevents cholesterol buildup in the arterial walls.
5. Chylomicron is responsible for carrying mostly triglycerides, and some cholesterol, from the intestines through the lymph system to the blood stream.

STUDENT WORKSHEET 4-1

1. Crab Cakes:
   a. Substitute whipped egg whites for part or all of mayonnaise.
   b. Use more herbs and spices to increase flavor.
   c. Sauté in a small amount of monounsaturated oil such as olive oil or canola oil or use a spray.
   d. Sauté in a nonstick pan.
   e. Serve with salsa to increase flavor.

2. Au Gratin Potatoes:
   a. Substitute the 2 tablespoons butter with a monounsaturated oil such as canola oil.
   b. Add garlic to onion for more flavor.
   c. Coat the baking pan with cooking spray instead of grease.
   d. Substitute low-fat or nonfat sour cream for the heavy cream.
   e. Use less Gruyere cheese and top the dish with some bread crumbs and paprika.
   f. Replace part of the cheese with an equal amount of skim mozzarella.
   g. Reduce the amount of cheese, then use some of the cheese to sprinkle on top toward the end of the cooking to bring out more flavor.
   h. Instead of making a sauce to pour over the potatoes, arrange half of the potato slices in the baking dish and then drizzle with sautéed onion, a small amount of salt and pepper, and half the cheese. Next, arrange the rest of the potato slices in the pan and repeat. Pour 1 cup heated skim milk over the potatoes and bake.

3. Blueberry Cobbler:
    a. Substitute margarine for part or all of the butter.
    b. Substitute 2 egg whites for the whole egg.
    c. Substitute ¼ cup reduced-fat milk for the milk.
    d. Substitute frozen vanilla yogurt for the vanilla ice cream.

STUDENT WORKSHEET 4-2 (Fat substitutes are in **bold**.)

Creme-Filled Chocolate Cupcakes—0 grams fat/cupcake
Sugar, **water**, corn syrup, bleached flour, egg whites, nonfat milk, defatted cocoa, invert sugar, **modified food starch (corn, tapioca)**, glycerine, fructose, calcium carbonate, natural and artificial flavors, leavening, salt, dextrose, calcium sulfate, **oat fiber, soy fiber**, preservatives, **agar**, sorbitan monostearate, mono- and diglycerides, **carob bean gum**, polysorbate 60, sodium stearoyl lactylate, **xanthan gum**, sodium phosphate, **maltodextrin**, **guar gum**, pectin, cream of tartar, sodium aluminum sulfate, artificial color.

Low-Fat Mayonnaise Dressing—1 gram fat per tablespoon
Water, corn syrup, liquid soybean oil, **modified food starch**, egg whites, vinegar, **maltodextrin**, salt, natural flavors, gums **(cellulose gel and gum, xanthan)**, artificial colors, sodium benzoate and calcium disodium EDTA.

Lite Italian Dressing—0.5 grams fat/2 tablespoons
Water, distilled vinegar, salt, sugar, contains less than 2% of garlic, onion, red bell pepper, spice, natural flavors, soybean oil, **xanthan gum**, sodium benzoate, potassium sorbate and calcium disodium EDTA, yellow 5 and red 40.

Light Cream Cheese—5 grams fat per 2 tablespoons
Pasteurized skim milk, milk, cream, contains less than 2% of cheese culture, sodium citrate, lactic acid, salt, stabilizers (**xanthan and/or carob bean and/or guar gums**), sorbic acid, natural flavor, vitamin A palmitate.

STUDENT WORKSHEET 4-3

Dairy Product	Kcalories/cup	Fat/cup	Sat Fat/cup	Cholesterol/cup
Whole Milk	146	8 grams	4.5 grams	24 milligrams
Reduced Fat Milk (2% fat)	122	4.8	3.1	20
Low Fat Milk (1% fat)	102	2.4	1.5	12
Fat-Free milk (skim)	83	0.2	0.2	5
Ice Cream, vanilla	266	14	9	58
Reduced fat Ice cream, vanilla	184	6	3.4	18
Frozen Yogurt, Vanilla	228	8	5	2

Which milk has the most fat? Whole milk

Which milk has the least fat? Fat-free or skim milk

Does fat-free milk contain less saturated fat than reduced-fat or regular milk? Yes

Does fat-free milk contain less cholesterol than reduced-fat or regular milk? Yes

How many kcalories do you save by drinking fat-free rather than regular milk? 63 kcal/cup

Which frozen dessert tends to be lowest in fat and saturated fat? frozen yogurt

STUDENT WORKSHEET 4-4

	Total Fat	Saturated Fat	Trans Fat	Mono Fat	Cholesterol
1 cup shortening	12.8	5.2	1.4 – 4.2	5.7	7
1 cup margarine-stick, 80% fat	11.0	2.1	2.8	5.2	0
2/3 cup soybean oil	9	1.3	0	2.1	0

Best nutritional profile: soybean oil cookie

Second - best nutritional profile: margarine

CHAPTER 5

Matching
1. e
2. h
3. k
4. m
5. p
6. c
7. f
8. j
9. h
10. q
11. a
12. g
13. o
14. l
15. d
16. b
17. i

True/False
1. T
2. F
3. F
4. T
5. F
6. T
7. F
8. T
9. T
10. F

Multiple Choice

1. b
2. c
3. d
4. d
5. a
6. a
7. a

Short Answer

1. The tertiary structure of protein-how it folds and twists-makes it able to perform its functions in the body.
2. The greatest amount of protein is needed when the body is building new tissues rapidly, such as during pregnancy or infancy.
3. Proteins act as taxicabs in the body, transporting iron and other minerals, some vitamins, fats, and oxygen through the blood.
4. Even though most plant proteins are examples of incomplete proteins they and other incomplete proteins are complementary proteins. Together they provide the necessary amino acids that are provided by complete proteins.
5. Marasmus
6. Antigens are foreign invaders in the body, and antibodies are proteins in the blood that bind to them.
7. Hormones are chemical messengers secreted into the bloodstream by various organs, such as the liver, to travel to a target organ and influence what it does.
8. Like carbohydrates and fats, proteins contain carbon, hydrogen and oxygen. Unlike carbohydrates and fats, proteins contain nitrogen and provide much of the body's nitrogen.

STUDENT WORKSHEET 5-2

1. RDA 68 g protein
2.

Foods and Beverages (including portion size)	Grams of Protein
Breakfast	
1-1/2 cups corn flakes	3
1 cup milk	8
1 cup orange juice	2
AM Snack	
3-1/2-inch bagel	7
2 tablespoons cream cheese	2
Lunch	
Double cheeseburger	21
Medium Coke	0
Medium French fries	6
PM Snack	
Apple – medium	0
Dinner	
Fried chicken, ¼ breast	18
1 cup mashed potatoes	4

1 cup canned corn	5
1 cup milk	8
Snack	
1 cup vanilla ice cream	4
Total Protein:	88

CHAPTER 6

Matching

1. e
2. g
3. i
4. c
5. m
6. l
7. o
8. f
9. d
10. h
11. a
12. k
13. n
14. p
15. j
16. b

Multiple Choice

1. b
2. c
3. c
4. d
5. b
6. b
7. a
8. b
9. b
10. b

True/False

1. T
2. F
3. F
4. T
5. T
6. F
7. F
8. T
9. F
10. F

Short Answer

1. Fat-soluble vitamins (A, D, E, and K)
2. Vitamin E
3. Vitamin D
4. Anencephaly
5. Choline

CHAPTER 7

Matching

1. m
2. d
3. i
4. o
5. f
6. p
7. c
8. k
9. n
10. l
11. b
12. j
13. g
14. e
15. a
16. h

Multiple Choice
1. a
2. b
3. a
4. b
5. d
6. c
7. a
8. c
9. d
10. c

True/False
1. F
2. T
3. T
4. F
5. F
6. T
7. F
8. T

Short Answer
1. Calcium and phosphorus
2. Electrolytes
3. Feeling tired and weak; decreased work performance; difficulty maintaining body temperature; and decreased immune function.
4. Zinc
5. Fluorosis

CHAPTER 8

Matching
1. e
2. g
3. i
4. f
5. k
6. o
7. c
8. m
9. p
10. a
11. h
12. l
13. j
14. d
15. n
16. b

Multiple Choice
1. c
2. a
3. d
4. e
5. a
6. d
7. c
8. a

True/False
1. T
2. T
3. T
4. F
5. T
6. F
7. T
8. T

Short Answer

1. Rosemary
2. Paprika
3. Turmeric
4. Reduction, Searing, Deglazing, Sweating, and Puréeing
5. They are made from the amino acids in protein when the foods are cooked at high temperatures.
6. They are cancer-causing substances produced from grilled food falling or hot coals or lava/ceramic bricks and carried to the food by smoke.
7. Stock
8. Seasonings are substances that bring out flavor that is already present. Flavorings add a new flavor or modify the original flavor of food or drink.

CHAPTER 9

Multiple Choice

1. b
2. e
3. c
4. e

True/False

1. T
2. F
3. F
4. T
5. T

Short Answer

1. Hypertension, type 2 diabetes, and obesity
2. Nonfat yogurt
3. Dijon mustard, shallots or garlic, and lemon or lime juice
4. Creamy dressings, puréed dressings, and reduction dressings
5. Too much heat and/or direct heat may toughen the cheese
6. The premise of catering breaks is to provide snacks that are satisfying and well-balanced so participants can keep their focus on the meeting they are attending.

CHAPTER 10

Matching

1. c
2. b
3. e
4. d
5. a

4. b
5. d
6. c
7. a
8. a
9. c

Multiple Choice

1. b
2. e
3. d

True/False

1. F
2. T
3. T
4. T
5. T
6. F
7. T
8. F

Short Answer

1. The major sources of added sweeteners in the diet come from soft drinks, candy and sugars, baked goods, fruit drinks, and dairy desserts and sweetened milk.
2. Fruits and vegetables; meat; milk-based items; and potatoes, rice corn and beans.
3. Hard cheeses, such as Swiss or Parmesan, contain very little lactose and don't usually cause symptoms because most lactose is removed during processing or is digested by the bacteria used in making cheese.

CHAPTER 11

Matching

1. m
2. j
3. e
4. o
5. d
6. i
7. b
8. k
9. q
10. n
11. h
12. a
13. l
14. f
15. c

Multiple Choice

1. a
2. c
3. a
4. b
5. b
6. d
7. e
8. d
9. b
10. a

True/False

1. T
2. F
3. T
4. F
5. F
6. T
7. T
8. F
9. T
10. T

Short Answer

1. A stroke will occur
2. Primary hypertension has an unknown cause, whereas secondary hypertension is persistently elevated blood pressured caused by a medical problem.

3. Systolic pressure, the top blood pressure number, measures the blood within the arteries when the heart is pumping. Diastolic pressure is the pressure in the arteries when the heart is resting between beats.
4. Phytochemicals
5. Blindness, dental disease, and nervous system disease.

CHAPTER 12

Matching

1. b
2. d
3. a
4. c

Multiple Choice

1. c
2. e
3. e
4. a
5. b
6. d
7. e
8. d
9. c
10. b

True/False

1. F
2. T
3. F
4. T
5. F
6. T
7. F
8. F
9. T
10. F

Short Answer

1. Body mass Index (BMI) is a method of measuring the degree of obesity. It is a direct calculation based on height and weight, and it applies to both men and women.
2. Increased self-image, increased stamina, and increased resistance to fatigue.
3. The maximum heart rate can be calculated by subtracting your age from 220.
4. No foods should be forbidden.
5. Gastric bypass is a popular malabsorptive operation in which a small stomach pouch is created to resist food intake. Next, a section of the small intestine is attached to the pouch to allow food to bypass the lower stomach and part of the upper small intestine. This reduces the amount of kcalories and nutrients absorbed by the body.

CHAPTER 13

Matching

1. k
2. n
3. b
4. g
5. j
6. d
7. p
8. f
9. l
10. a
11. i
12. m
13. c
14. e
15. o
16. h

Multiple Choice

1. b
2. b
3. b
4. a
5. d
6. b
7. c
8. d
9. c
10. b

True/False

1. T
2. F
3. F
4. T
5. F
6. T
7. F
8. T
9. T
10. T

Short Answer

1. 2–4 pounds
2. Get an early start, delay artificial nipples, and watch for infections.
3. 1,000 kcalories
4. 4–6 pounds per year
5. 4–5 years of age